Conceptual Breakthroughs in Evolutionary Genetics

Conceptual Breakthroughs in Evolutionary Genetics
A Brief History of Shifting Paradigms

John C. Avise

AMSTERDAM • BOSTON • HEIDELBERG • LONDON
NEW YORK • OXFORD • PARIS • SAN DIEGO
SAN FRANCISCO • SINGAPORE • SYDNEY • TOKYO
Academic Press is an imprint of Elsevier

Academic Press is an imprint of Elsevier
525 B Street, Suite 1800, San Diego, CA 92101-4495, USA
32 Jamestown Road, London NW1 7BY, UK
225 Wyman Street, Waltham, MA 02451, USA

Notice
No responsibility is assumed by the publisher for any injury and/or damage to
persons, or property as a matter of products liability, negligence or otherwise, or from
any use or, operation of any methods, products, instructions or ideas contained in the
material herein. Because of rapid advances in the medical sciences, in particular,
independent verification of diagnoses and drug dosages should be made.

British Library Cataloguing-in-Publication Data
A catalogue record for this book is available from the British Library

Library of Congress Cataloging-in-Publication Data
A catalog record for this book is available from the Library of Congress

ISBN: 978-0-12-420166-8

For information on all Academic Press publications
visit our website at elsevierdirect.com

Printed and bound by CPI Group (UK) Ltd, Croydon, CR0 4YY

Contents

Part II
The Next 50 Years (1910–1960): Expanding the Foundations

Part IV
Post-1980: Elaborating and Revisiting
the Foundations

Acknowledgments

Francisco Ayala and Andrei Tatarenkov kindly made helpful comments on early drafts of the manuscript. Several anonymous reviewers made superb suggestions that improved the presentation. The author's work is supported by funds from the University of California at Irvine.

The study of the history of a field is the best way of acquiring an understanding of its concepts.

Ernst Mayr (1982, p. 20)

It is helpful for both students and professional biologists if the concepts and theories that form the basis for our understanding are put into the context of how the field has developed over time. ... It is also arguably more fun to learn things if we understand why particular problems have stimulated the curiosity of biologists or how technical advances made new discoveries possible.

Tobias Uller (2013, p. 267)

Science is ill suited for people who demand eternal truths. Scientific lessons preached by one generation often seem doomed to be modified if not overturned by the generations that follow. All active sciences are ever changing, but perhaps nowhere has this been certifiably truer than in the field of evolutionary genetics. Hardly a month now goes by without the announcement of some substantial discovery in evolutionary biology or genetics that forces us to rethink what we thought we knew about nature's operations. The grandest scientific breakthroughs may bring fame (such as Nobel Prizes) to their authors, but lesser findings are important too because their cumulative effects may alter the trajectory of a discipline.

In 1962, the philosopher of science Thomas Samuel Kuhn published a book — *The Structure of Scientific Revolutions* — that itself promoted a revolutionary notion: science progresses not in a continuous linear fashion but rather via "paradigm shifts" in which the scientific community — in response to contrarian evidence and a shifting consensus of opinion — eventually abandons conventional wisdom and replaces it with a competing account of reality. Each such paradigm shift constitutes a "scientific revolution". How well the Kuhnian worldview generally characterizes science is debatable. It certainly applies in some cases, but many other keystone discoveries (such as the Nobel Prize-winning invention of the polymerase chain reaction [PCR], or the elucidation of DNA's double-helical structure) seem to be singular scientific breakthroughs that did not overturn any particular conventional wisdom on the topic. Regardless of exactly how science unfolds in practice, change it does, often quite inexorably. What was taught to me as a graduate student in the early 1970s would bear scant resemblance to a comparable program of advanced evolutionary genetics in the year 2010. For those who worship entrenched scientific dogma above all else, paradigms lost in science may indeed conjure images of John Milton's (1667) *Paradise Lost.*

This book highlights 70 conceptual paradigms in evolutionary biology and genetics that were challenged (and in many cases subsequently abandoned or modified) following landmark discoveries that superseded scientists' prior understandings of nature. Not all of these paradigm shifts conform strictly to the Kuhnian model, but each does represent a major alteration of scientific attitudes before versus after the path-breaking finding(s). In this book, the 70 paradigm shifts are arranged chronologically by their approximate or exact date of occurrence. Each short essay is presented in the following format: the standard paradigm at that time (i.e., the paradigm that would be lost); the ensuing conceptual revolution; a subjective "paradigm-shift index" (*PS-score*) followed by a brief explanation; and a few key references and suggestions for further reading (arranged chronologically) from the relevant literature.

The *PS-score* can range from 1 to 10, with higher numbers indicating greater overall impact on the field. Thus, each assigned *PS-score* represents my attempt to integrate several considerations: the revolutionary nature of the new scientific worldview; its veracity (the strength and durability of its supporting evidence); the temporal duration of its impact to date (thus biasing in favor of earlier discoveries and biasing against more recent ones, all else being equal); and the scientific breadth of the new paradigm's implications. By assigning a less than perfect score to a given paradigm shift, in no way do I mean to belittle its importance — all of the conceptual conversions discussed in this book rank among the most profound insights in evolutionary genetics over the past one-and-a-half centuries, but some surely were more consequential than others.

I wrestled with several alternative ways to categorize the conceptual breakthroughs so as to embrace the great heterogeneity among the discoveries. For example, some of them were revolutionary ideas for explaining longstanding observations (e.g., *Whose Handiwork?*, *Organelle Origins* [Chapters 1 and 38, respectively]); others were startling empirical discoveries for their time (e.g., *Genetic Variation*, *Split Genes*, *Regulatory RNAs* [Chapters 37, 49, 63]); others were striking innovations (e.g., *Molecular Phylogeny*, *Kin Selection* [Chapters 31, 32]); and others were novel ideas about a circumscribed issue (e.g., *Sex Ratio*, *Aging* [Chapters 18, 26]). In another way of categorizing the paradigm shifts, some of the breakthroughs were driven by one single-minded genius (such as Lynn Margulis in *Organelle Origins*, Willi Hennig in *Cladistics*, or Barbara McClintock in *Jumping Genes* [Chapters 38, 35 and 24, respectively]); some were the result of many people racing one another toward a common goal (e.g., *Genetic Material*, *Genomic Sequencing* [Chapters 23, 66]); and some represent a clarification of thought or a re-awakening of an existing idea (e.g., *Individual Selection*, *Exaptations*, *Adaptive Speciation* [Chapters 36, 52, 69]). I invite readers to formulate their own classification scheme for the 70 breakthroughs. Such an exercise might be particularly interesting for historians or philosophers of science.

In the end, I divided this book into four parts based on chronology. Part I deals with discoveries in the first 50 years following Darwin (1859–1910); Part II treats various breakthroughs over the next 50 years (1910–1960); Part III covers the 1960s and 1970s; and Part IV covers the modern era, post-1980.

I will not be too disappointed if this book provokes or even irritates some readers, because different evolutionary geneticists inevitably will have differing opinions about what merits inclusion as a salient paradigm shift in the field. I have tried to be inclusive in my choice of topics and discoveries, but for readers that remain disgruntled, all I can suggest is that they themselves try to add, subtract, or differently rank various scientific discoveries from the list. Nevertheless, I do apologize to any practitioner who may feel that his or her own paradigm-busting work has been unduly neglected.

This pithy book is intended to be fun and educational for a wide audience of biologists and science historians. In addition to summarizing the chronology and many major developments in the field of evolutionary genetics, I hope that this treatment may inspire a new generation of students and practitioners to challenge conventional scientific wisdom (critically but responsibly), and perhaps even strive to formulate and rank new research paradigms of their own. In science, a fine line sometimes exists between being a true visionary and a rabble-rousing dissident, but successfully navigating that intellectual tightrope can be a richly rewarding experience.

REFERENCES

Milton J. 1667. *Paradise Lost*. Simmons, London, UK.

Kuhn T. 1962. *The Structure of Scientific Revolutions*. University of Chicago Press, Chicago, IL.

Mayr E. 1982. *The Growth of Biological Thought*. Harvard University Press, Cambridge, MA.

Uller T. 2013. Non-genetic inheritance and evolution. In: Kampourakis K. (Ed.). *The Philosophy of Biology: A Companion for Educators*. Springer, New York, NY, pp. 267–287.

The First 50 Years (1859–1910): Laying the Foundations

The breakthroughs in Part I begin with Charles Darwin's treatises on natural selection (1859) and sexual selection (1871), and Gregor Mendel's (1865) unanticipated documentation of particulate inheritance. They continue with several other highly notable accomplishments, including the births of biogeography (1876) and biochemical genetics (1902), and they conclude with Morgan's (1910) key discoveries that linked the observed behaviors of chromosomes to Mendel's inferred laws of genetic segregation and independent assortment. In these and in the dozen other discoveries included in Part I, biologists thereby began to lay the conceptual foundations for what later would become recognizable as the field of evolutionary genetics.

Chapter 1

1859
Whose Handiwork?

5   10

THE STANDARD PARADIGM

Life's beauty and variety are prima facie evidence of God's creative handiwork.
Throughout most of human history, and including parts of classical Greek
philosophy to 19th century theology, a standard sentiment was that supernatural
agents (Gods) were directly responsible for the diversity and the exquisite
functional features of life. The 17th century naturalist John Ray famously
referred to this notion as "The Wisdom of God". In 1802, this traditional
paradigm – natural theology – was again epitomized in a powerful treatise by
the Anglican minister William Paley, who paid homage to a beneficent Creator
God for directly fashioning organic material into the myriad phenotypic features
that enable organisms to survive and reproduce. In the modern era, natural
theology has resurfaced in the guise of the "intelligent design" (ID) religious
movement. Proponents of ID posit that complex biological outcomes, ranging
from bacterial cells to human beings, were purposefully designed and directly
crafted by a supreme intelligence (e.g., by a Creator God) rather than having
arisen via non-sentient natural evolutionary forces.

The dilemma for natural theology had always been the fact that many
biological phenomena seem inconsistent with a wise and benevolent Creator.
Why, for example, do parasites and diseases exist, why are pain and suffering
so prevalent, and, more generally, why are biological imperfections and the
natural equivalent of "evil" apparently so common in the organic world? This
perennial philosophical problem is known as *theodicy* (literally translated as
"God's justice", from the Greek words *theós* for God and *dik* for justice).
The word theodicy was coined in 1710 by the German Philosopher and
mathematician Gottfried Leibniz in an essay entitled *Theodicic Essays on the
Benevolence of God, the Free Will of Man, and the Origin of Evil.*

J.C. Avise: Conceptual Breakthroughs in Evolutionary Genetics.
DOI: http://dx.doi.org/10.1016/B978-0-12-420166-8.00001-4

3

THE CONCEPTUAL REVOLUTION

Charles Darwin (and, independently, Alfred Russel Wallace) identified a non-sentient evolutionary process — natural selection — that can sculpt organic material (forge adaptations) without the need to invoke either conscious or direct supernatural causation. Darwin adduced that evolution is a necessary consequence of natural selection as the explanation of biological design because natural selection fosters the adaptation of organisms to their environmental milieu. Natural selection not only provided a powerful explanation for evolution, but thereby also offered a potential solution to the longstanding theodicic conundrum. Natural selection is a mindless process of nature, as unthinking and uncaring as gravity or lightning. No longer need rationalists directly blame a Creator God for life's miseries (or directly credit Him for biological adaptations). Instead, the proximate driving force of evolution is an unconscious and non-omnipotent agent, which obviates the necessity to invoke supernatural agents to account for life's flaws as well as beauties. The concept of biological evolution by natural selection — and *ergo* the influence of Darwinian evolutionary thought — has spread far beyond biology and well into the realms of the social sciences, philosophy, and religion.

PS-score: 10

This perfect score is merited by the fact that scientists ever since Darwin have universally accepted the reality of evolution by natural selection (among many other population genetic forces). In terms of epistemological significance, the Darwinian revolution of the 19th century — by unshackling science from untestable religious doctrines — did for biology what the Copernican revolution of the 16th century had done for the physical sciences. The Darwinian conceptual revolution has transformed and unified thought throughout the biological sciences. Indeed, it has been aptly stated that nothing in biology makes sense except in evolution's light (Dobzhansky, 1973).

REFERENCES AND FURTHER READING

Ray J. 1691. *The Wisdom of God Manifested in the Works of Creation*. Smith, London, UK.

Paley W. 1802. *Natural Theology*. Reprinted by Oxford University Press, Oxford, UK, 2006.

Darwin C. 1859. *On the Origin of Species*. John Murray, London, UK.

Dobzhansky T. 1973. Nothing in biology makes sense except in the light of evolution. *Am. Biol. Teacher* 35:125–129.

Dawkins R. 1987. *The Blind Watchmaker*. Norton & Co., New York, NY.

Ruse M. 1999. *The Darwinian Revolution*. University of Chicago Press, Chicago, IL.

Ruse M. 2003. *Darwin and Design*. Harvard University Press, Cambridge, MA.

Manning RE (Ed.). 2013. *The Oxford Handbook of Natural Theology*. Oxford University Press, Oxford, UK.

Chapter 2

1861 Spontaneous Generation

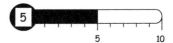

THE STANDARD PARADIGM

Life continually arises spontaneously. A widespread belief from the time of the ancient Romans through and beyond the Middle Ages was that organic life routinely generates from non-life, such as when rats emerge from a heap of trash, amphibians appear each spring from swampy mud, or maggots swarm from rotting meat. This standard folklore is known as abiogenesis, or the origin of life via spontaneous generation.

THE CONCEPTUAL REVOLUTION

The revised worldview is that only life normally begets life. The "disproof" of spontaneous generation traditionally is ascribed to the French chemist and microbiologist Louis Pasteur, who in 1861 conducted a critical laboratory experiment showing that sterilized broth cultures do not regrow microorganisms unless exposed to suitable inocula containing other microbes. Apparently, the microbes were proliferating rather than arising spontaneously.

PS-score: 5

Pasteur's experiments — important though they were — were not entirely novel in concept, because two centuries earlier the Italian physician Francesco Redi similarly had disproved the maggots-from-meat hypothesis simply by keeping adult flies (and thus their eggs) away from rotting meat. Thus, Pasteur merely extended doubts about spontaneous generation into the microbial realm. Also, Pasteur's work gets only a modest *PS-score* because, by hard criteria, his experiments did not prove that life has never arisen from inorganic materials, but rather that any such spontaneous process apparently

J.C. Avise: *Conceptual Breakthroughs in Evolutionary Genetics.*
DOI: http://dx.doi.org/10.1016/B978-0-12-420166-8.00002-6

does not happen routinely, even in the microbial world. Nearly a century later, other kinds of experiments (see Chapter 27) would suggest that organic life has indeed arisen from inorganic substrates on at least one occasion in the Earth's long geological history.

FURTHER READING

Levine R, Evers C. 1999. *The Slow Death of Spontaneous Generation (1668–1859).* National Health Museum, Washington, DC.

Chapter 3

1865
The Nature of
Heredity

THE STANDARD PARADIGM

Inheritance is analogous to the mixing of fluids, such that offspring show a smooth blend of miscible characteristics transmitted from their parents. From routine crosses of domesticated animals and plants, as well as from personal human family experiences, people throughout the ages had always accepted what seemed obvious: that some form of blending inheritance (such as a "mixing of the bloods") must account for the general trend toward offspring intermediacy in phenotypic traits (such as body shape or facial features) between the sire and dam. Whatever fluid-like medium might govern heredity seemed to be thoroughly blendable during offspring production.

THE CONCEPTUAL REVOLUTION

By tallying numbers of offspring displaying alternative traits in experimental crosses involving true-breeding inbred strains of pea plants, Gregor Mendel deduced the particulate nature of inheritance. The non-miscible particles that Mendel uncovered would later — in 1909 — be named "genes" (see Chapter 12). Mendel's findings revealed two of the most fundamental rules of heredity for diploid organisms: (1) the law of segregation, which states that the two alleles (alternative forms of a gene) segregate from one another during gametogenesis (the production of gametes); and (2) the law of independent assortment, which states that alleles at separate loci normally segregate independently of one another during gamete formation. Later discoveries (see Chapters 13 and 14) would identify exceptions to both of these Mendelian laws, but such exceptions did more to highlight the generality of Mendel's rules than to dishonor them.

J.C. Avise: Conceptual Breakthroughs in Evolutionary Genetics.
DOI: http://dx.doi.org/10.1016/B978-0-12-420166-8.00003-8

PS-score: 10

This perfect score is merited by the fact that Mendel's discoveries about heredity have withstood the test of time, proved generalizable to all kinds of multicellular organisms, and become a solid foundation for the entire field of genetics. Indeed, if inheritance truly were miscible rather than particulate, the population genetic variation that is prerequisite for evolution would be rapidly lost (halved) in each successive generation of sexual reproduction. But Mendel in effect showed that genetic variation could be maintained indefinitely in populations because the particles of heredity tend to maintain their separate identities across the generations. Mendel's elucidation of the particulate nature of inheritance ranks second on the all-time list of the most important conceptual breakthroughs in evolutionary genetics, trailing only Darwin's elucidation of natural selection as the principal shaping force of adaptive evolution (see Chapter 1).

REFERENCE

Mendel GJ. 1865. Versuche über pflanzenhybriden [Experiments on plant hybridization]. *Verhandlungen des Naturforschenden Vereins (Bruenn)* 4:3—47.

Chapter 4

1871
Pre-copulatory
Sexual Selection

THE STANDARD PARADIGM

Only humans exercise refined mate choice. In the pre-Darwinian era most people assumed that non-human animals were too unsophisticated to exert deliberate choice about with whom to mate. Although beasts mated almost exclusively with members of their own kind (i.e., conspecifics), any further refinement in mating preferences was presumed to be far beyond their behavioral wherewithal.

THE CONCEPTUAL REVOLUTION

In 1871, Charles Darwin greatly elaborated on a powerful concept — sexual selection — that he had introduced only briefly in *The Origin*. According to Darwin, the struggle for existence involves both differential survival and reproduction. In *The Origin*, Darwin focused primarily on natural selection via differential survival, whereas in the *Descent of Man and Selection in Relation to Sex* he focused on differential reproduction via sexual selection, which he subdivided into two major categories: male—male combat and female choice. In male—male combat, males fight or otherwise compete against one another ultimately for sexual access to females, with the net result being sexual selection favoring the evolution of male-specific features such as sharp leg spurs on roosters or large antler racks on buck deer. In female choice (also known as epigamic selection), females preferentially mate with males that display particular phenotypic features such as colorful plumages and melodious songs in male birds, fancy fins of male guppies, or the showy and perfumed flowers of plants that attract pollinators. Whereas natural selection forges adaptations to the environment, sexual selection in effect forges phenotypes that contribute directly to mating success. Sexual

J.C. Avise: Conceptual Breakthroughs in Evolutionary Genetics.
DOI: http://dx.doi.org/10.1016/B978-0-12-420166-8.00004-X

selection often operates in opposition to natural selection, because a trait that attracts a mate might also be disadvantageous with respect to promoting individual survival. Thus, sexual selection can account for the origin and maintenance of many phenotypic features that are difficult to explain as adaptations to the environment.

PS-score: 10

Sexual selection is a major shaping force of evolution, perhaps second only to natural selection (see Chapter 1). Its elucidation revolutionized thought about evolutionary processes and made evolutionary sense of countless phenotypic features that otherwise might seem non-adaptive or even maladaptive. In recent decades, the concept of sexual selection has been greatly elaborated and also extended to encompass post-copulatory sperm competition and cryptic gametic choice within the female reproductive tract (see Chapter 43).

REFERENCES AND FURTHER READING

Darwin C. 1871. *The Descent of Man and Selection in Relation to Sex.* John Murray, London.

Cronin H. 1991. *The Ant and the Peacock: Altruism and Sexual Selection from Darwin to Today.* Cambridge University Press, Cambridge, UK.

Zuk M. 2002. *Sexual Selections: What We Can and Can't Learn about Sex from Animals.* University of California Press, Berkeley, CA.

Reznick DN. 2010. *The Origin Then and Now: An Interpretive Guide to the Origin of Species.* Princeton University Press, Princeton, NJ.

Ruse M. (Ed.). 2013. *The Cambridge Encyclopedia of Darwin and Evolutionary Thought.* Cambridge University Press, Cambridge, UK.

Chapter 5

1875
Nature versus
Nurture

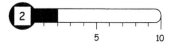

THE STANDARD PARADIGM

The etiologies of most human traits are impossible to tease apart. Although genetic factors and rearing conditions probably can explain variation in many human attributes, there seemed to be no ready way to distinguish hereditary from environmental influences on quantitative traits such as various behavioral phenotypes.

THE CONCEPTUAL REVOLUTION

In the late 1860s and early 1870s, the Victorian polymath Francis Galton (a cousin of Charles Darwin) drew attention to the use of twin studies for elucidating the relative importance of hereditary factors and environmental rearing conditions in features such as human intelligence. Although the modern concept and cellular basis of monozygotic twinning had not yet been established, Galton's basic idea was that because many human twins are presumably genetically identical, any observed differences between such individuals must be due to something (namely, rearing conditions) other than their genes. Galton was a remarkable individual who is also credited with pioneering the statistical concept of correlation, and founding such diverse fields as meteorology (scientific weather forecasting), forensics (using human fingerprints), eugenics (which envisions the genetic improvement of humanity), and psychometrics (the science of measuring human mental faculties). His scientific legacies are thus varied and profound.

Throughout much of the ensuing 20th century and beyond, three types of twin protocols tracing back to Galton have been employed in scientific investigations of genetic versus cultural influences on human behaviors. The least critical approach analyzes trait correlations in monozygotic twins reared

J.C. Avise: Conceptual Breakthroughs in Evolutionary Genetics.
DOI: http://dx.doi.org/10.1016/B978-0-12-420166-8.00005-1

11

apart (MZA twins), the rationale being that any resemblance between MZA twins suggests (but does not prove) genetic influence. The second approach compares trait correlations for monozygotic twins versus dizygotic (non-identical or fraternal) twins, under the rationale that any higher correlations observed in identical twins register genetic influences. The third approach compares monozygotic twin-sets reared together versus those reared apart, the rationale being that any greater trait differences between separated twins must register environmental effects. Each of these approaches has conceptual as well as logistical limitations, so none yields definitive conclusions about the relative impacts of "hardwired" versus environmental influences on particular phenotypic features.

PS-score: 2

Twin studies helped to initiate and propagate the longstanding "nature versus nurture" controversy in evolutionary genetics, and thus were of considerable historical importance to the field. However, the modern paradigm is that both genes *and* the environment, as well as interactions between the two, are all important contributors to phenotypic variation in most quantitative traits (in humans as well as in other creatures).

REFERENCES AND FURTHER READING

Galton F. 1869. *Hereditary Genius*. Macmillan, London, UK.

Wilson EO. 1978. *On Human Nature*. Harvard University Press, Cambridge, MA.

Lewontin RC, Rose SW, Kamin LJ. 1984. *Not in Our Genes*. Random House, New York, NY.

Bouchard TJ, Lykken DT, McGue M, Segal NL, Tellegen A. 1990. Sources of human psychological differences: the Minnesota study of twins reared apart. *Science* 250:223–228.

Avise JC. 1998. *The Genetic Gods: Evolution and Belief in Human Affairs*. Harvard University Press, Cambridge, MA.

Burbridge D. 2001. Francis Galton on twins, heredity and social class. *Br. J. Hist. Sci.* 34:323–340.

Chapter 6

1876
Biogeography

THE STANDARD PARADIGM

Species arise in situ either by special creation or by natural selection. In the 1800s few European naturalists had travelled extensively outside of Europe, and thus they lacked much personal experience about the profound impact of geography on the planet's distributions of animals and plants. A dramatic exception was the British naturalist and explorer Alfred Russel Wallace — the co-discoverer of natural selection (see Chapter 1) — who at the age of 25 mounted the first of his two major scientific expeditions to distant lands: first to the Brazilian Amazon where he spent four years collecting biological specimens; and later to southeastern Asia and the Malay Archipelago where he did likewise for another eight years. On these remarkable journeys, Wallace documented how biotas can vary dramatically from place to place as a consequence of their particular evolutionary histories and biogeographical relationships.

THE CONCEPTUAL REVOLUTION

In his extensive natural-history writings that included 20 books and about 800 articles recounting his travel adventures, Wallace (1855) promulgated the concept that (p. 196) "every species has come into existence coincident both in space and time with a preexisting closely allied species". He thus added the key dimension of historical biogeography to evolutionary discussions of species' origins and organismal distributions. His pioneering tome on this subject was published in 1876.

PS-score: 10

Alfred Russel Wallace is universally recognized as the founding father of biogeography — the study of the geographical distributions of organisms.

J.C. Avise: Conceptual Breakthroughs in Evolutionary Genetics.
DOI: http://dx.doi.org/10.1016/B978-0-12-420166-8.00006-3

13

Even today, the field of biogeography remains one of the sturdiest pillars supporting the entire pantheon of evolutionary biology (see Chapter 51). Wallace's writings contained many biogeographic insights that might today seem almost so obvious as to be taken for granted. To pick just four of many examples, Wallace noted that large taxonomic groups (such as classes and orders) are generally spread over large sections of the Earth whereas smaller ones (such as families and genera) are frequently confined to particular portions; no group or species has come into existence twice; many well-marked groups of species are found in closely adjoining localities; and the distribution of the organic world across time is very similar to its present distribution in space.

REFERENCES AND FURTHER READING

Wallace AR. 1855. On the law which has regulated the introduction of new species. *Annals Magazine on Natural History* 16:184—196.

Wallace AR. 1876. *The Geographical Distribution of Animals*. Harper & Brothers, New York, NY.

Andrewartha HG, Birch LC. 1954. *The Distribution and Abundance of Animals*. University of Chicago Press, Chicago, IL.

Fichman M. 2004. *An Elusive Victorian: The Evolution of Alfred Russel Wallace*. University of Chicago Press, Chicago, IL.

Brown JH, Lomolino MV. 2008. *Biogeography*, 2nd edition. Sinauer, Sunderland, MA.

Smith CH, Beccaloni G. (Eds). 2008. *Natural Selection and Beyond: The Intellectual Legacy of Alfred Russel Wallace*. Oxford University Press, Oxford, UK.

Berry A. 2013. Evolution's red-hot radical. *Nature* 496:162—164.

Chapter 7

1889
Germ Plasm

THE STANDARD PARADIGM

Body cells contribute hereditary particles to the next generation. Darwin called these imagined particles gemmules, and he proposed that they were routinely shed by somatic tissues and carried by the bloodstream to the reproductive organs, where they accumulated in the sex cells for transmission to progeny. If true, this "theory of pangenesis" could imply that phenotypic changes in the soma acquired during the lifetime of an individual might be passed on to offspring.

THE CONCEPTUAL REVOLUTION

The German evolutionary biologist August Weismann introduced a germ plasm theory in which he posited that inheritance is exclusively a function associated with germ cells (sperm and ova in animals) and that the soma plays no role in this process. Weismann's proposal thereby made a clean distinction between the germline and the soma, such that whereas germ cells produce somatic cells the latter do not feed hereditary information back into the production of gametes. This genetic demarcation between germline cells and somatic cells became known as the Weismann barrier.

PS-score: 9

Weismann's ideas were important because they seemed to contradict widely discussed proposals by the French evolutionary biologist Jean-Baptiste Lamarck that phenotypic features acquired during the lifetimes of parents would be genetically transmitted to progeny. Today, the distinction between germline and soma is a central hereditary paradigm. Furthermore, Lamarckian modes of inheritance are generally dismissed (but see Chapter 22) in favor of strict Mendelian genetic mechanisms (see Chapter 3).

J.C. Avise: *Conceptual Breakthroughs in Evolutionary Genetics.*
DOI: http://dx.doi.org/10.1016/B978-0-12-420166-8.00007-5
15

REFERENCES AND FURTHER READING

Lamarck J-B. 1809. *Philosophie Zoologique: ou Exposition des Considérations Relatives à l'Histoire Naturelle des Animaux [Zoological Philosophy: Exposition with Regard to the Natural History of Animals]*. Natural History Museum, Paris, France.

Weismann A. 1889. *Essays Upon Heredity*. Clarendon Press, Oxford, UK.

Churchill FB. 1968. August Weismann and a break from tradition. *J. Hist. Biol.* 1:91–112.

Mayr E. 1982. *The Growth of Biological Thought*. Harvard University Press, Cambridge, MA.

Chapter 8

1902
The Etiology of
Disabilities

THE STANDARD PARADIGM

Human ailments and disabilities are caused by exogenous (and often super-natural) agents such as evil spirits or bad karma. Across the millennia, this standard dogma has been the motivation for countless ameliorative practices ranging from fervent religious prayers to herbal medicines to witchdoctors' séances and incantations.

THE CONCEPTUAL REVOLUTION

Based on his research conducted near the turn of the 20th century, the English physician Archibald Garrod reported discoveries that fundamentally challenged conventional wisdom by demonstrating that four rather common human conditions (alkaptonuria, albinism, cystinurea, and pentosuria) result from endogenous bodily foul-ups that proved to be both heritable and mecha-nistically understandable in simple genetic (Mendelian) and biochemical terms. For example, we now know that alkaptonuria is caused by a genetic defect in the metabolic pathway for phenylalanine and tyrosine (two of the 20 amino acids from which cells construct proteins), and that this can lead to degenerative arthritis in the large joints and spine. Garrod christened such outcomes "inborn errors of metabolism". In other words, many (but certainly not all) biological diseases could be ascribed to mis-workings of the human metabolic machinery, which itself was governed at least in part by heritable genetic factors that obey Mendelian principles. More broadly, Garrod's genetic findings regarding the etiology of human genetic diseases also extended scientific knowledge about the particulate nature of inheritance from Mendel's pea plants (see Chapter 3) to *Homo sapiens*.

J.C. Avise: Conceptual Breakthroughs in Evolutionary Genetics.
DOI: http://dx.doi.org/10.1016/B978-0-12-420166-8.00008-7

PS-score: 9

Today, many thousands of inborn metabolic disorders are documented to be underlain by specifiable genetic lesions or mutations in various among the legions of "housekeeping" and other functional genetic loci in the human genome (the full suite of genetic material within each somatic or germline cell). Extensive computer databases of these genetic disorders are maintained and updated routinely. Fittingly, Garrod is remembered as the founding father of human biochemical genetics and metabolomics.

REFERENCES AND FURTHER READING

Garrod AE. 1902. The incidence of alkaptonuria: a study in chemical individuality. *Lancet* ii:1616–1620.

Garrod AE. 1923. *Inborn Errors of Metabolism*, 2nd edition. [1st edition published in 1909]. Frowde, Hodder & Stoughton, London, UK.

Bearn AG. 1993. *Archibald Garrod and the Individuality of Man*. Oxford University Press, New York, NY.

Scriver CR, Beaudet WS, Sly WS, Valle D. (Eds). 2001. *The Metabolic and Molecular Basis of Inherited Disease*. 8th edition. McGraw-Hill, New York, NY.

Nyhan WL, Barstop B, Ozand PT. 2005. *Atlas of Metabolic Diseases*, 2nd edition. Oxford University Press, Oxford, UK.

Avise JC. 2010. *Inside the Human Genome: A Case for Non-intelligent Design*. Oxford University Press, Oxford, UK.

Chapter 9

1902
Autosomes

THE STANDARD PARADIGM

The physical basis of heredity remained obscure. Although Mendelian particles must exist (see Chapter 3), where are they housed within cells? Following the rediscovery of Mendel's work in 1900 (see Chapter 11), questions arose as to where such entities (genes) might reside and exactly how Mendelian laws might be physically underpinned. The answers soon would be forthcoming.

THE CONCEPTUAL REVOLUTION

In 1902, Walter Sutton and Theodor Boveri independently discerned the link between Mendel's abstract hereditary factors and tangible structures known as chromosomes that cytologists had discovered in the late 1800s. Sutton's work — based on his Master's thesis — entailed cytological observations on the large chromosomes of the landlubber grasshopper (*Brachystola magna*). Under a microscope, Sutton watched as the pairs of chromosomes segregated and sorted independently during gamete formation, in a nicely parallel fashion to the deduced behavior of Mendelian particles. Sutton (1902) was crystal clear in his paper's understated final sentence (p. 39): "I may finally call attention to the probability that the association of paternal and maternal chromosomes in pairs and their subsequent separation during the reducing division ... may constitute the physical basis of the Mendelian law of heredity." This cytological work on autosomes (nuclear chromosomes other than the sex chromosomes) was soon confirmed and extended in the laboratory of Thomas Hunt Morgan (see Chapter 14), and it later led to what E. B. Wilson (1925) termed the "Sutton-Boveri" chromosomal theory of inheritance.

J.C. Avise: Conceptual Breakthroughs in Evolutionary Genetics.
DOI: http://dx.doi.org/10.1016/B978-0-12-420166-8.00009-9

PS-score: 7

This breakthrough merits a high score because it provided the needed link between deduced Mendelian particles and the physical structures (chromosomes) on which they reside. The chromosomal theory of inheritance has fully withstood the tests of time. Today, we know that chromosomes are the physical structures that house genes in essentially all organisms, and that their cellular behaviors during meiosis and syngamy basically account for Mendel's laws of inheritance in sexually reproducing taxa.

In an interesting footnote to this story, the term "genome" (see Chapter 66) apparently traces to a merger in 1920 of "gene" with "chromosome" by the German botanist Hans Winkler (Lederberg and McCray, 2001).

REFERENCES AND FURTHER READING

Sutton WS. 1902. On the morphology of the chromosome group in *Brachystola magna. Biol. Bull.* 4:24–39.

Sutton WS. 1903. The chromosomes of heredity. *Biol. Bull.* 4:231–251.

Wilson EB. 1925. *The Cell in Development and Heredity*, 3rd edition. Macmillan, New York, NY.

Lederberg J, McCray AT. 2001. 'Ome sweet 'omics: a genealogical treasury of words. *The Scientist* 15:8.

Crow EW, Crow JF. 2002. Walter Sutton and the chromosome theory of heredity. *Genetics* 160:1–4.

Chapter 10

1905
Epistasis

THE STANDARD PARADIGM

Genes act independently to influence organismal traits. Although Gregor Mendel (see Chapter 3) studied some "dihybrid" crosses (those involving two genes) in his experiments on pea plants, the particular loci that he monitored proved to be independent in their effects on various of the plants' phenotypes. Thus, the only genetic interactions that he observed involved intra-locus effects, as for example when one allele was dominant or recessive to its counterpart at the same diploid gene (see Chapter 11).

THE CONCEPTUAL REVOLUTION

Genes sometimes interact with one another to determine an organism's phenotype. This discovery came about early in the 20th century when the English geneticist William Bateson and his colleagues conducted experimental crosses on chickens with different comb characteristics, and pea plants with different flower colors. By counting progeny displaying various phenotypic conditions, the researchers deduced that two or more genes could interact in non-additive ways to affect a given phenotypic outcome. Any such gene-by-gene interaction became known as epistasis, a phenomenon that is now appreciated to be ubiquitous in the biological world. For example, epistasis might occur when a genotype at one locus conceals the phenotypic effects of alleles at another gene. Thus, epistasis is an intergenic interaction somewhat analogous to the phenomena of dominance and recessivity at the intragenic (inter-allelic) level.

PS-score: 6

Bateson was an ardent champion of Gregor Mendel's work, and he also coined the term *genetics*. Epistasis does not contradict Mendelian principles but rather embellishes them in ways reflective of the reality that separate

J.C. Avise: Conceptual Breakthroughs in Evolutionary Genetics.
DOI: http://dx.doi.org/10.1016/B978-0-12-420166-8.00010-5

genes often interact non-linearly and non-additively to underpin organismal traits. Today, epistasis is widely appreciated to be the norm rather than the exception for many phenotypic characteristics ranging from biochemical machinations to morphological features. Indeed, given the interconnectivity of complex metabolic networks (and their genetical underpinnings), it would be highly surprising if epistasis as well as pleiotropy (when a gene simultaneously influences multiple phenotypic features) were not common genetic phenomena in the organic world.

REFERENCES AND FURTHER READING

Bateson W. 1902. *Mendel's Principles of Heredity: A Defence*. Cambridge University Press, Cambridge, UK.
Bateson W. 1913. *Problems of Genetics*. Yale University Press, New Haven, CT.
Bateson P. 2002. William Bateson: A biologist ahead of his time. *J. Genet*. 81:49–58.

Chapter 11

1908
Hardy-Weinberg

THE STANDARD PARADIGM

Genetic dominance and recessivity imply something about allele frequencies in populations. Mendel's empirical findings about inheritance (see Chapter 3) went unnoticed by the scientific community until they were rediscovered independently by Hugo de Vries, Carl Correns, and Erich Tschermak in 1900. Among Mendel's observations was the fact that some alleles appear to be "dominant" whereas others are "recessive" with respect to their influence on organismal phenotype. In the early 1900s, some biologists and statisticians naïvely interpreted these words to mean that dominant alleles would reach high frequencies in populations and recessive alleles would remain rare. This was a blatant misconstrual of terms; Mendel's concepts of dominance and recessivity referred strictly to allelic effects on the phenotype.

THE CONCEPTUAL REVOLUTION

In 1908, the English mathematician Godfrey Hardy and the German physician Wilhelm Weinberg independently straightened out this latter-day confusion by demonstrating what subsequently became known as the Hardy-Weinberg law. Consider alleles a and b, with frequencies p and q in a population, such that $p + q = 1$. Assuming random mating in a sexual population of large size, the frequencies of diploid genotypes are given by a binomial expansion of the sum of the allele frequencies: $(p + q)^2 = p^2 + 2pq + q^2$, corresponding to aa, ab, and bb. Furthermore, in the absence of counteracting evolutionary forces (such as natural selection or recurrent mutation), these represent equilibrium frequencies that are stably maintained generation after generation.

PS-score: 8

The Hardy-Weinberg law is fundamental to the field of population genetics (the study of genetic changes in populations through time). Hardy-Weinberg

J.C. Avise: *Conceptual Breakthroughs in Evolutionary Genetics.*
DOI: http://dx.doi.org/10.1016/B978-0-12-420166-8.00011-7
23

equilibrium gives the expected population genetic configuration of diploid genotypes in the absence of disturbing evolutionary forces, in analogous fashion to how Newton's first law in mechanics states that a body remains at rest or maintains a constant velocity when not acted upon by external agents. Particular kinds of departures from the expectations of Hardy-Weinberg law or Newton's law can then be highly informative about additional evolutionary factors or physical forces, respectively, that might be causing any observed digressions from the anticipated equilibrium states. For example, if a population displays an excess of heterozygotes relative to expectations under Hardy-Weinberg equilibrium (HWE), then this might suggest that heterozygous genotypes have a higher genetic fitness than homozygotes.

REFERENCES AND FURTHER READING

Dunn LC. 1965. *A Short History of Genetics*. McGraw-Hill, New York, NY.

Hardy GH. 1908. Mendelian proportions in a mixed population. *Science* 28:49–50.

Weinberg W. 1908. Über dem nachweis der vererbung beim menschen. *Ver. Vaterl. Naturk. Württemberg* 64:369–382. [English translation in Boyer SH. 1963. *Papers on Human Genetics*. Prentice-Hall, Englewood Cliffs, NJ.]

Punnett RC. 1950. Early days of genetics. *Heredity* 4:1–10.

Ayala FJ, Avise JC (Eds). 2013. *Essential Readings in Evolutionary Biology*. Johns Hopkins University Press, Baltimore, MD.

Chapter 12

1909
Genotype versus
Phenotype

THE STANDARD PARADIGM

Hereditary principles need clarification. Although the work of Gregor Mendel (see Chapter 3) had been rediscovered by the scientific community in 1900 (see Chapter 11), confusion still reigned about the nature of Mendelian particles and how they act during an individual's development and over the course of a population's evolution.

THE CONCEPTUAL REVOLUTION

A breakthrough of sorts occurred in 1909 when the Danish botanist/geneticist Wilhelm Johannsen introduced three simple but concept-laden words: "gene", "genotype", and "phenotype". In particular, Johannsen drew a distinction between the hereditary dispositions of organisms (their genotypes) and the manners in which those dispositions manifest themselves in an organism's physical appearances (phenotypes). In between the genotype and the phenotype resides the full suite of evolutionary developmental processes (see Chapter 22) that influence how each genotype translates during ontogeny into an organism's physical and behavioral makeup.

Johannsen's distinction of genotype from phenotype clarified what had been a source of confusion between an organism's hereditary potential and its realization. It also bore some resemblance to August Weismann's distinction between the germline and the soma (see Chapter 7), in the sense that both authors recognized that causal connections were unidirectional (see Chapter 42) between an organism's hereditary constitution and its emergent physical attributes. Both authors thereby also contributed to the wane of Lamarckian ideas on the inheritance of acquired characteristics.

J.C. Avise: Conceptual Breakthroughs in Evolutionary Genetics.
DOI: http://dx.doi.org/10.1016/B978-0-12-420166-8.00012-9

25

PS-score: 7

Today, the terms gene, genotype, and phenotype continue to be universally applied. Indeed, it is hard to imagine a substantive conversation in evolutionary genetics that did not somehow reference these seminal concepts. The terms have also been further clarified and elaborated. For example, depending on the context, the word phenotype can be applied to any trait or set of features ranging from the biochemical (such as the precise operation of a metabolic pathway) to the level of an organism's gross anatomy (such as the presence versus absence of wings or feathers). The word genotype is likewise catholic and — depending on the context — can apply at levels ranging from the allelic constitution of a particular genetic locus to the composite makeup of a multi-locus assemblage of genes.

REFERENCES AND FURTHER READING

Johannsen WL. 1909. Elemente der Exakten Erblichkeitslehre [*The Elements of an Exact Theory of Heredity*]. Gustav Fisher, Jena, Germany.

Johannsen WL. 1911. The genotype conception of heredity. *Amer. Natur.* 45:129–159.

Chapter 13

1909
Non-Mendelian
Inheritance

THE STANDARD PARADIGM

Genes in higher organisms tend to obey Mendel's laws of segregation and independent assortment. (Actually, however, this paradigm had existed only for the few years following the rediscovery of Mendel's work at the turn of the century, because at that time Mendelian principles had been documented in only a few organisms such as pea plants [see Chapter 3] and humans [see Chapter 8].)

THE CONCEPTUAL REVOLUTION

Carl Correns and Erwin Baur, working independently, were the first to describe a hereditary mode (in this case involving plant plastids) that did not conform to Mendel's rules. These scientists had uncovered examples of maternal inheritance wherein traits are transmitted to seedlings through the mother only, with the pollen-providing parent apparently having no influence on the phenotype of the progeny. The authors correctly concluded that in addition to Mendelian genes in the cell nucleus, additional hereditary factors must occur in the cellular cytoplasm that comes mostly from the egg. Today, we understand that such hereditary factors reside in the chloroplast genomes (cpDNA) of plants and in the mitochondrial genomes (mtDNA) of animals and plants (see Chapter 38).

PS-score: 5

In the full century following Correns and Baur, many more categories and countless examples of non-Mendelian heredity in eukaryotic organisms have

J.C. Avise: Conceptual Breakthroughs in Evolutionary Genetics.
DOI: http://dx.doi.org/10.1016/B978-0-12-420166-8.00013-0

come to light. For example, gene conversion is a type of non-Mendelian heredity in which one allele at a locus converts (via mismatch repair) to the DNA sequence of the other allele at that diploid locus. Such gene conversion can be considered one category of meiotic drive or molecular drive wherein one allele out-competes the other for transmission to the next organismal generation (see Chapter 48). Somewhat analogously, genetic recombination events between loci are also known, and these sometimes can lead to a form of "concerted" evolution in which paralogous members of a gene family tend to evolve in concert within a species, with respect to their nucleotide sequences.

Notwithstanding the near-universality of non-Mendelian maternal inheritance for cytoplasmic genomes, and the diversity of other non-Mendelian hereditary mechanisms subsequently uncovered, the fact remains that Mendel's laws as originally formulated remain applicable to a vast majority of nuclear loci in most eukaryotic species. Thus, Mendelian heredity remains the rule rather than the exception, accounting for why I have given the discovery of non-Mendelian inheritance only an intermediate *PS-score*.

REFERENCES AND FURTHER READING

Baur E. 1909. The nature and the inheritance properties of horticultural varieties of *Pelargonium zonale* having white borders [original title was in German]. *Zeitschr. f. ind. Abst. u. Vererbungsl* 1:300–351.

Correns C. 1909. Inheritance experiments with pale (yellow) green and variegated varieties of *Mirabilis julapa* [original title was in German]. *Zeitschr. f. ind. Abst. u. Vererbungsl* 1:291–329.

Dover GA. 1982. Molecular drive: a cohesive mode of species evolution. *Nature* 299:111–117.

Liao D. 1999. Concerted evolution: molecular mechanism and biological implications. *Amer. J. Human Genet.* 64:24–30.

Hagemann R. 2000. Erwin Baur and Carl Correns: Who really created the theory of plastid inheritance? *J. Heredity* 91:435–440.

Chen JM, Cooper DN, Chuzhanova N, Férec C, Patrinos GP. 2007. Gene conversion: mechanisms, evolution, and human disease. *Nature Rev. Genet.* 8:762–775.

Chapter 14

1910
Sex Chromosomes

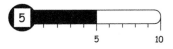

5   10

THE STANDARD PARADIGM

Genes are inherited equally through both parents in sexual species. Although a few examples of maternal inheritance had been reported (see Chapter 13), the broader sentiment at that time was that genes typically are transmitted to progeny from both parents via Mendel's recently rediscovered laws of heredity (see Chapter 11). What remained uncertain were the physical composition of Mendelian particles, and how and precisely where such genes are housed in cells (but see Chapter 9).

THE CONCEPTUAL REVOLUTION

In 1908, Thomas Hunt Morgan established a *Drosophila* (fruit fly) laboratory at Columbia University (New York) to which he recruited several soon-to-be famous graduate students and postdocs, including Alfred Sturtevant, Hermann Muller, C.B. Bridges, and Theodosius Dobzhansky. During the 1910s and 1920s, this "fly lab" made several landmark discoveries that eventuated in Morgan's receipt of the 1933 Nobel Prize in Physiology or Medicine for his body of work on hereditary transmission mechanisms in *Drosophila*. Morgan's lab confirmed the suspicions that genes are sequentially housed along chromosomes, that each chromosome contains a set of linked genes with partially non-independent transmission from one generation to the next, and that the closeness of genetic linkage quantitatively predicts departures from Mendel's law of independent assortment for unlinked loci.

Morgan also discovered sex chromosomes and the phenomenon of sex-limited inheritance. In *Drosophila* (as in humans, mammals, and many other organisms), females carry two X chromosomes whereas males carry one X and one Y chromosome. Females transmit one copy of the X to each son and daughter, and males pass their copy of the X to daughters and the Y to sons. Genes housed on these sex chromosomes thus have a sex-linked mode of inheritance

J.C. Avise: Conceptual Breakthroughs in Evolutionary Genetics.
DOI: http://dx.doi.org/10.1016/B978-0-12-420166-8.00014-2

29

that differs from the standard pattern of Mendelian inheritance as applied to unlinked loci on the autosomes (chromosomes that are not sex-linked). Many years later other categories of sex linkage were uncovered, such as the ZW system of birds and butterflies in which females have the sex-chromosomal constitution ZW and males are ZZ.

PS-score: 5

The discovery of sex linkage was important because it (1) revealed taxonomically widespread departures from standard autosomal inheritance, and (2) initiated critical thought about the optimal evolutionary strategies of genes that are transmitted and expressed differently in males and females (see Chapter 48).

REFERENCES AND FURTHER READING

Morgan TH. 1910. Sex limited inheritance in *Drosophila*. *Science* 32:120–122.

Bridges CB. 1925. Sex in relation to chromosomes and genes. *Amer. Natur.* 59:127–137.

Charlesworth B. 1991. The evolution of sex chromosomes. *Science* 251:1030–1033.

Steinman M, Steinman S, Lottspeich F. 1993. How Y chromosomes become inert. *Proc. Natl. Acad. Sci. USA* 9:5737–5741.

Rice WR. 1996. Evolution of the Y sex chromosome in animals. *BioScience* 46:331–343.

Bainbridge D. 2003. *The X in Sex: How the X Chromosome Controls Our Lives*. Harvard University Press, Cambridge, MA.

Jones S. 2003. *The Descent of Men: Revealing the Mysteries of Maleness*. Houghton Mifflin, Boston, MA.

The Next 50 Years (1910–1960): Expanding the Foundations

The conceptual breakthroughs in Part II encompass a broad spectrum of discoveries that were central to the development of the so-called Modern Synthesis in evolutionary biology. Nearly all of these discoveries came in the "pre-molecular" era of evolutionary genetics, and indeed it was not until nearly mid-century that scientists finally confirmed that nucleic acids (rather than proteins) are the hereditary material of life. The first half of the 20th century also witnessed the birth and growth of population genetics, further characterization of the nature of mutations, elaboration of the biological species concept, the discovery of transposable elements, and a host of other scientific developments whose legacies are so fully infused into modern evolutionary genetics as to now seem like self-evident truths. Such is the nature of progress in science, where ideas that once were radical become thoroughly incorporated into what we later take for granted as scientific knowledge.

Chapter 15

1912
Continental Drift

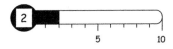

THE STANDARD PARADIGM

The Earth's landmasses and their respective biotas had always been in the current configuration. It seemed inconceivable to most scientists that cartographic features of the planet had changed much across evolutionary time.

THE CONCEPTUAL REVOLUTION

In 1912, the German geophysicist/meteorologist Alfred Wegener published a provocative article speculating that the planet's continents had formerly been united (into a supercontinent Pangea) before they splintered and slowly plowed across the Earth's surface to their present-day positions. This hypothesis became known as Wegener's theory of continental drift. Wegener backed his idea — which got a hostile reception at the time — by several lines of empirical evidence: continental shapes that fit together like pieces of a jigsaw puzzle; particular resemblances between the biotas on apposite landmasses, such as those on alternate sides of the Atlantic Ocean (some of these similarities previously had been attributed to hypothetical land bridges); and a matching alignment of geological formations on the respective landmasses, such as similar rock strata in the Appalachian Mountains of North America and the Scottish Highlands. Wegener's theory languished for several decades until — in the 1950s and 1960s — it received compelling support from detailed geophysical explorations of the Earth's crustal movements (plate tectonics). Today, continental drift is widely accepted as a geological fact beyond dispute, and the precise rate of modern crustal movements (typically several centimeters per year) can even be monitored directly using satellite and other global-positioning technologies.

J.C. Avise: Conceptual Breakthroughs in Evolutionary Genetics.
DOI: http://dx.doi.org/10.1016/B978-0-12-420166-8.00015-4

PS-score: 2

Wegener's ideas eventually were to revolutionize thought in geology and paleontology. They also dovetailed nicely with some of the biological concepts that had been implicit in various of Wallace's biogeographic principles (see Chapter 6). In evolutionary genetics, Wegener's insights understandably had a lesser impact, except insofar as they helped to promote the broader realization that biogeographic patterns and processes (see Chapter 51) must take into account the surprisingly dynamical nature of the Earth's historical geology and physiography.

REFERENCES AND FURTHER READING

Wegener AL. 1912. Die entstehung der kontinente [The formation of the continents]. *Geologische Rundschau* 3:276–292.

Wegener AL. 1968. *The Origin of Continents and Oceans* [translated from a later edition of a book originally published in German in 1915]. Mathuen, London, UK.

Chapter 16

1915
Homeotic Genes

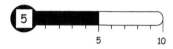

THE STANDARD PARADIGM

Animal body plans are conservative. Although genes were known that could substantially modify organismal phenotypes (such as Mendel's short versus tall pea plants), it was widely assumed that the body parts of conspecific animals were structurally conservative during the evolutionary short term. However, little was known about the genetic underpinnings of organismal development (ontogeny), or how ontogenetic shifts might relate to evolutionary processes.

THE CONCEPTUAL REVOLUTION

In 1915, Calvin Bridges (see Bridges and Morgan, 1923) uncovered the first example of a "homeotic" mutation, which he named *bithorax* (*bx*). All adult insects have three thoracic segments (T1, T2, and T3) each normally bearing a pair of legs, with T2 and T3 also supporting wings and flight organs called halteres, respectively. Fruit flies homozygous for Bridges' *bx* mutation developed as if T3 had been transformed into T2, such that individuals had an extra pair of wings (for a total of four rather than the usual two). Other homeotic mutations soon were discovered that acted during development in such a way as to transform (for example) antennae into legs that sprouted from a fly's head!

Today, we know that homeotic genes encode factors that regulate gene transcription and that they occur as tight collinear clusters of about a dozen loci (derived through serial gene duplications) within the nuclear genomes of invertebrate and vertebrate animals. During an organism's development, these "*Hox*" genes govern somatic differentiation along the primary anterposterior body axis and they also play key regulatory roles in the construction of secondary body axes such as limbs. Thus, *Hox* genes are of exceptional importance in directing morphogenesis during ontogeny, and they also are

J.C. Avise: Conceptual Breakthroughs in Evolutionary Genetics.
DOI: http://dx.doi.org/10.1016/B978-0-12-420166-8.00016-6

35

presumed to underlie many of the alterations in animal body plans during the evolutionary process.

PS-score: 5

The discovery of homeotic genes left an interesting scientific legacy. On the one hand, it fed into the Mendelian−Biometrician controversy of the early 1900s (see Chapter 19) by raising the possibility that evolution sometimes proceeds by saltational jumps in phenotype, and it thereby also gave fodder to Richard Goldschmidt's (1940) suggestion that "hopeful monsters" might play a significant evolutionary role. These latter notions are no longer popular today. On the more positive side of the ledger, the discovery of homeotic genes helped to lay the foundation for a related concept − now widely accepted − that changes in gene regulation generally underlie the evolutionary diversification of organismal phenotypes (see Chapter 41). Furthermore, the study of homeotic genes and numerous other categories of regulatory loci has become a solid cornerstone of much modern research in "evo-devo" (evolutionary developmental biology).

REFERENCES AND FURTHER READING

Bridges CB, Morgan TH. 1923. *The third-chromosome group of mutant characters of* Drosophila melanogaster. The Carnegie Institute, Washington, DC.

Goldschmidt R. 1940. *The Material Basis of Evolution.* Yale University Press, New Haven, CT.

Carroll SB. 1995. Homeotic genes and the evolution of arthropods and chordates. *Nature* 376:479−485.

Carroll SB, Grenier JK, Weatherbee SD. 2001. *From DNA to Diversity: Molecular Genetics and the Evolution of Animal Design.* Blackwell, Malden, MA.

Carroll SB. 2005. *Endless Forms Most Beautiful: The New Science of Evo Devo and the Making of the Animal Kingdom.* Norton, New York, NY.

Hoffer A, Xiang J, Pick L. 2013. Variation and constraint in *Hox* gene evolution. *Proc. Natl Acad. Sci. USA* 110:2211−2216.

Chapter 17

1927
Mutation

THE STANDARD PARADIGM

Mutations are extremely rare phenotypic aberrations of unknown etiology. In 1901 the Dutch botanist/geneticist Hugo de Vries introduced the word "mutation" to describe spontaneous alterations in the hereditary material in the *Oenothera* primrose plants on which he worked. By the early 1900s mutations were understood to underlie organic evolution, but they remained ill defined and poorly characterized. That situation gradually improved as researchers in the "fly lab" of Thomas Hunt Morgan at Columbia University (see Chapter 14) began to dissect the chromosomal nature of heredity using experiments involving the transmission genetics of mutations underlying various aberrant phenotypes in fruit flies. One of those researchers was Hermann J. Muller.

THE CONCEPTUAL REVOLUTION

In their genetic mapping efforts, Muller and his colleagues had been hampered by the sluggish rate of naturally occurring mutations, which meant that their research was constrained by the available number of mutant phenotypes. In 1927, Muller published an article in which he documented that artificially exposing fruit flies to heavy doses of X-rays "caused a rise of about fifteen thousand per cent" in the frequencies of mutations induced in germline cells (see Chapter 7). Muller concluded: "there can be no doubt that many, at least, of the changes produced by X-rays are of just the same kind as the 'gene mutations' which are obtained, with so much greater rarity, without such treatment, and which we believe furnish the building blocks of evolution." In 1946, Muller received a Nobel Prize for his discovery of mutagenic agents and their phenotypic effects.

J.C. Avise: Conceptual Breakthroughs in Evolutionary Genetics.
DOI: http://dx.doi.org/10.1016/B978-0-12-420166-8.00017-8

PS-score: 3

Muller's experimental breakthrough had several consequences: it speeded up the genetic mapping experiments on fruit flies, which contributed to an understanding of the chromosomal basis of heredity (see Chapter 9); it helped uncover the etiology and phenotypic consequences of *de novo* mutations; and it left Muller and many other geneticists with the distinct impression that mutations generally are harmful and impose a "genetic load" on the populations that harbor them. This sentiment in turn contributed to the classical school of thought about the evolutionary significance of genetic variability (see Chapter 37). After World War II and the atomic bombings of Hiroshima and Nagasaki in Japan, Muller involved himself in political campaigns about the biological dangers of radiation and atomic testing in the atmosphere.

REFERENCES AND FURTHER READING

Muller HJ. 1927. Artificial transmutation of the gene. *Science* 66:84–87.

Muller HJ. 1950. Our load of mutations. *Amer. J. Human Genet.* 2:111–176.

Muller HJ. 1962. *Studies in Genetics.* Indiana University Press, Bloomington, IN.

Sturtevant AH. 1965. *A History of Genetics.* Harper & Row, New York, NY.

Carlson EA. 1981. *Genes, Radiation, and Society: The Life and Work of H.J. Muller.* Cornell University Press, Ithaca, NY.

Wallace B. 1991. *Fifty Years of Genetic Load.* Cornell University Press, Ithaca, NY.

Neel JV. 1994. *Physician to the Gene Pool.* John Wiley & Sons, New York, NY.

Chapter 18

1930
Sex Ratio

THE STANDARD PARADIGM

Group advantage accounts for sex ratios (numbers of males versus females) in populations. For example, Darwin speculated that a 1:1 sex ratio in a population might minimize overall competition over mates, whereas other evolutionary biologists reasoned that a male-biased sex ratio would benefit a species because each male can sexually service many females, and still others argued that a female-biased sex ratio would be ideal because this would maximize a population's reproductive output.

THE CONCEPTUAL REVOLUTION

Ronald Fisher was one of the three leading architects of theoretical population genetics during the early to mid-20th century (see Chapter 19). He had profound insights on many population topics, not the least being on sex ratios. Fisher reasoned as follows. Every generation of individuals, regardless of the population sex ratio, contains an equal number of autosomal alleles inherited from male and female parents. This fundamental truth led him to realize that a form of frequency-dependent selection operates in such a way as to favor parental investment strategies that culminate in approximately equal sex ratios. If males are infrequent in a population, families producing disproportionate numbers of sons (relative to their costs of production) will on average leave more grandchildren than do families that produce excess daughters, so any genes for male-producing tendencies temporarily spread. Conversely, if females are infrequent in a population, families producing disproportionate numbers of daughters will on average leave more grandchildren than do families that produce excess sons, so any genes for female production will temporarily tend to spread. In other words, autosomal genes tend to maximize their mean fitness by producing whichever sex is in the minority, such that at equilibrium a 1:1 sex ratio in the population typically is approached. Over the

J.C. Avise: Conceptual Breakthroughs in Evolutionary Genetics.
DOI: http://dx.doi.org/10.1016/B978-0-12-420166-8.00018-X

course of evolution, this selectively favored equilibrium (from the perspective of autosomal genes) has been genetically codified in many animal species by the evolution of sex-determining chromosomes (see Chapter 14) and by a meiotic process that assures a more-or-less equal production of male-yielding and female-yielding gametes.

In recent years, numerous examples of highly unequal sex ratios have also come to light, often involving strong excesses of females in particular species. Accordingly, much research has shifted to a focus on the evolutionary and mechanistic explanations – such as selection operating on female-transmitted genes and intracellular parasites housed in the cellular cytoplasm – for such gender-biased sex ratios.

PS-score: 2

Fisher's sex-allocation concept was revolutionary within its domain, but it had limited broader impact on evolutionary genetics, except insofar as it contributed to a growing realization that naïve "group selectionism" is an outmoded train of thought and should be replaced by evolutionary scenarios invoking selection operating at the fitness levels of genes (see Chapter 48), individuals (see Chapter 36), and families (see Chapter 32). Another intellectual legacy from the type of selective argumentation that Fisher fostered includes an eventual rise of the field of evolutionary ecology (which weds evolutionary reasoning to population demography and ecology).

REFERENCES AND FURTHER READING

Fisher RA. 1930. *The Genetical Theory of Natural Selection*. Clarendon Press, Oxford, UK.

MacArthur RH. 1961. Population effects of natural selection. *Amer. Natur.* 95:195–199.

Emlen JM. 1973. *Ecology: An Evolutionary Approach*. Addison-Wesley, Reading, MA.

Pianka ER. 1974. *Evolutionary Ecology*. Harper & Row, New York, NY.

Charnov EL. 1982. *The Theory of Sex Allocation*. Princeton University Press, Princeton, NJ.

O'Neill SL, Hoffmann AA, Werren JH (Eds). 1997. *Influential Passengers: Inherited Microorganisms and Arthropod Reproduction*. Oxford University Press, Oxford, UK.

Hardy ICW (Ed.). 2002. *Sex Ratios: Concepts and Research Methods*. Cambridge University Press, Cambridge, MA.

Majerus MEN. 2003. *Sex Wars: Genes, Bacteria, and Biased Sex Ratios*. Princeton University Press, Princeton, NJ.

Anderson TR. 2013. *The Life of David Lack: Father of Evolutionary Ecology*. Oxford University Press, Oxford, UK.

Chapter 19

1932
End of a Debate

THE STANDARD PARADIGM

Mendelian principles seemed to be incompatible with the gradual and cumulative mode of Darwinian evolution. During the late 1800s and early 1900s, some geneticists (the so-called Mendelians) accepted the universality of Mendel's rules of particulate inheritance but they thought this implied that evolution must proceed by discrete jumps or saltational changes (for example, from short to tall pea plants). Other scientists (the Biometricians) interpreted this evolutionary genetic conundrum to imply that many genes (such as those underlying human height and other continuously varying characters) might not fully obey Mendel's principles.

THE CONCEPTUAL REVOLUTION

In a remarkable burst of scientific creativity in the early 1930s, Ronald Fisher, J.B.S. Haldane, and Sewall Wright effectively silenced the debate between the Mendelians and the Biometricians by demonstrating, mathematically, that Mendelian inheritance is fully compatible with gradual Darwinian evolution. The key realization was that particulate Mendelian genes at multiple loci can in principle act in a multifactorial way to produce continuous population variation and gradual evolution in quantitative traits. Generally, the greater the number of genetic loci that cumulatively affect a trait, the more nearly continuous will be the phenotypic variation in that trait. Of course, various environmental influences might also impact a given quantitative character, thus potentially contributing further to continuous population variation in that phenotype.

J.C. Avise: Conceptual Breakthroughs in Evolutionary Genetics.
DOI: http://dx.doi.org/10.1016/B978-0-12-420166-8.00019-1

PS-score: 9

The classic mathematical works by Fisher, Haldane and Wright were revolutionary because, for the first time, they united the formerly separate realms of Mendelian heredity and Darwinian evolution into one intellectually consistent whole. They also laid the conceptual foundations for the fields of population genetics and what would become modern quantitative genetics (see also Chapter 5). Also, the mathematical constructs of population genetics—laid down by Fisher, Haldane, and Wright—would soon be married to ecology and organismal biology in a grand conceptual orchestration that would become known as the modern evolutionary synthesis (see Chapter 21).

REFERENCES AND FURTHER READING

Fisher RA. 1930. *The Genetical Theory of Natural Selection*. Clarendon Press, Oxford. UK.

Wright S. 1931. Evolution in Mendelian populations. *Genetics* 16:97−159.

Haldane JBS. 1932. *The Causes of Evolution*. Longmans & Green, London, UK.

Provine WB. 1971. *The Origins of Theoretical Population Genetics*. University of Chicago Press, Chicago, IL.

Provine WB. 1986. *Sewall Wright and Evolutionary Biology*. University of Chicago Press, Chicago, IL.

Chapter 20

1935
Biological Species

THE STANDARD PARADIGM

Species are biological discontinuities sufficiently well earmarked as to warrant formal recognition by taxonomists. Although Darwin provided a powerful solution to the longstanding question of how new species might arise, the biological category "species" remained somewhat ill defined in a broad conceptual sense. Indeed, Darwin viewed species merely as well-marked varieties falling along a biological continuum with conspecific populations.

THE CONCEPTUAL REVOLUTION

In the early 1900s, this situation changed as evolutionary biologists began to formulate more refined species definitions. Especially noteworthy was a seminal paper by Theodosius Dobzhansky in 1935, in which this Russian-born evolutionary biologist eloquently articulated what would later be called the biological species concept (BSC). Under this powerful notion, which applies mostly to sexually reproducing organisms, a biological species is a group of actually or potentially interbreeding populations reproductively isolated from other such groups (Mayr, 1963). Reproductive isolation was the key concept that gave a biological rationale for why the organic world appears to be subdivided into rather discrete packets called species.

PS-score: 6

This new paradigm was extremely important but less than revolutionary, because earlier authors had expressed similar sentiments and even Darwin's ideas can be read as generally consistent with the BSC. Alfred Russell Wallace, the co-discoverer of natural selection, likewise had anticipated the BSC when he wrote, in 1865, "Species are merely those strongly marked races or local forms which, when in contact, do not intermix, and when inhabiting

J.C. Avise: Conceptual Breakthroughs in Evolutionary Genetics.
DOI: http://dx.doi.org/10.1016/B978-0-12-420166-8.00020-8

distinct areas are generally believed ... to be incapable of producing a fertile hybrid offspring." In recent decades, many other species concepts have been introduced as alternatives to the BSC or (more often) as being complementary to it. For example, the phylogenetic species concept (PSC) emphasizes the historical or phylogenetic distinctiveness of populations as a basis for species recognition. But any such historical distinctiveness ultimately is promoted by reproductive isolating barriers that are the hallmark of the BSC.

REFERENCES AND FURTHER READING

Wallace AR. 1865. On the phenomenon of variation and geographical distribution as illustrated by the *Papilionidae* of the Malayan region. *Trans. Linn. Soc. Lond.* 25:1–71.

Dobzhansky T. 1935. A critique of the species concept in biology. *Phil. Sci.* 2:344–355.

Mayr E. 1963. *Animal Species and Evolution*. Harvard University Press, Cambridge, MA.

Cracraft J. 1983. Species concepts and speciation analysis. In: Johnson RF (Ed.). *Current Ornithology*. Plenum Press, New York, NY, pp. 159–187.

Wheeler QD, Meier R (Eds). 2000. *Species Concepts and Phylogenetic Theory*. Columbia University Press, New York, NY.

Hey J. 2001. *Genes, Categories, and Species*. Oxford University Press, New York, NY.

Coyne JA, Orr HA. 2004. *Speciation*. Sinauer, Sunderland, MA.

Price T. 2008. *Speciation in Birds*. Roberts & Co., Greenwood Village, CO.

Chapter 21

1937
Modern Synthesis

THE STANDARD PARADIGM

The field of evolutionary genetics is steeped in mathematical language. In the early 1930s, the concepts of Darwinian selection and Mendelian heredity were amalgamated into a grand quantitative construct of population genetics by the theoreticians Ronald Fisher, J.B.S. Haldane, and Sewall Wright (see Chapter 19). However, these mathematical treatments were somewhat inaccessible to many biologists. They needed translation via direct connections to empirical observations on ecological and evolutionary genetic processes.

THE CONCEPTUAL REVOLUTION

The first such biological translation occurred in 1937 with the publication of *Genetics and the Origin of Species* by the Russian-born geneticist Theodosius Dobzhansky. In the ensuing dozen years, the empirical side of evolutionary biology would be elaborated and extended in seminal books by the German systematist Ernst Mayr, the American paleontologist George Gaylord Simpson, and the American botanist/geneticist G. Ledyard Stebbins. Collectively, this body of work, when coupled to the earlier mathematical treatments, would become known as the modern evolutionary synthesis, or simply the "Modern Synthesis".

PS-score: 9

There can be no doubt that this biological synthesis represents one of the great achievements in the history of evolutionary genetics. This accomplishment is also of interest because it illustrates how major advances in biology sometimes differ from those in other hard sciences such as physics. For example, scientific breakthroughs in theoretical physics often seem to arise quite suddenly when singular mathematical formulations (often by

J.C. Avise: Conceptual Breakthroughs in Evolutionary Genetics.
DOI: http://dx.doi.org/10.1016/B978-0-12-420166-8.00021-X

rather young intellects) challenge conventional wisdom. In biology, by contrast, many of the grand achievements represent synthetic treatments (often published in book-length formats) summarizing a lifetime of accumulated evidence and ideas that unite broad swaths of data and theory. The Modern Synthesis provides a classic example of this latter brand of scientific revolution.

REFERENCES AND FURTHER READING

Dobzhansky T. 1937. *Genetics and the Origin of Species*. Columbia University Press, New York, NY.

Mayr E. 1942. *Systematics and the Origin of Species*. Columbia University Press, New York, NY.

Simpson GG. 1944. *Tempo and Mode in Evolution*. Columbia University Press, New York, NY.

Stebbins GL. 1950. *Variation and Evolution in Plants*. Columbia University Press, New York, NY.

Maynard Smith J. 1989. *Evolutionary Genetics*. Oxford University Press, Oxford, UK.

Mayr E. 2004. *What Makes Biology Unique? Considerations on the Autonomy of a Scientific Discipline*. Cambridge University Press, Cambridge, UK.

Fox CW, Wolf JB (Eds). 2006. *Evolutionary Genetics: Concepts and Case Studies*. Oxford University Press, Oxford, UK.

Chapter 22

1942
Epigenetics

THE STANDARD PARADIGM

Genes dictate phenotypes rather directly. Although the exact molecules of heredity were not yet firmly established in 1942 (see Chapter 23), a widespread supposition was that genetic loci (whatever they might prove to be) mechanistically program organismal phenotypes in relatively straightforward ways (see Chapter 12).

THE CONCEPTUAL REVOLUTION

Following general sentiments presaged as early as 1896 by James Baldwin, in 1942 Conrad Waddington coined the term epigenetics to refer to "the branch of biology which studies the causal interactions between genes and their products which bring the phenotype into being". Waddington was emphasizing the now-indisputable fact that between the genotype and the phenotype resides a whole suite of "epigenetic" developmental processes, including gene-by-gene interactions (see Chapter 10). These include the entire complex of cellular mechanisms by which genes are regulated, modulated, and expressed during ontogeny (see Chapter 41). Across the ensuing decades, the epigenetic concept has morphed into a variety of related postulates and nuanced viewpoints that remain under active scientific investigation. For example, one modern definition is that epigenetics is the study of alterations in gene expression or phenotype caused by cellular mechanisms other than changes in the underlying DNA sequences. Another modern definition places much greater emphasis on epigenetic developmental influences that actually are heritable across the generations (e.g., via maternal effects). Some of the more reformist epigenetic proposals even border on Lamarckian-like concepts (see Chapter 7), whereby phenotypic features acquired during the lifetime of a parent are passed on to its progeny. In recent years, there has been

J.C. Avise: Conceptual Breakthroughs in Evolutionary Genetics.
DOI: http://dx.doi.org/10.1016/B978-0-12-420166-8.00022-1

a resurgence of interest in the multitudinous mechanisms underlying various epigenetic phenomena and their broader relevance to evolutionary theory.

PS-score: 6

There is no question that Waddington's focus on developmental processes (ontogeny) during the evolutionary process was itself a timely development that has been passed on to several succeeding generations of scientists. The main reason I have not given this paradigm a higher *PS-score* is that this long-maturing epigenetic revolution is still underway, so it is still too early to know how its various perspectives will play out in terms of their overall impact on evolutionary genetic thought going forward. Today, many evolutionary geneticists are just as interested in the "epigenome" as they are in the genome *per se*.

REFERENCES AND FURTHER READING

Baldwin JM. 1896. A new factor in evolution. *Amer. Natur.* 30:441−451.

Waddington CH. 1942. The epigenotype. *Endeavour* (1942) 1:18−20.

Jablonka E, Lamb MJ. 1996. *Epigenetic Inheritance and Evolution: The Lamarckian Dimension.* Oxford University Press, New York, NY.

Riggs AD, Russo VEA, Martienssen RA. 1996. *Epigenetic Mechanisms of Gene Regulation.* Cold Spring Harbor Laboratory Press, Plainview, NY.

West-Eberhard MJ. 2002. *Developmental Plasticity and Evolution.* Oxford University Press, New York, NY.

Carey N. 2012. *The Epigenetics Revolution.* Columbia University Press, New York, NY.

Baylin S, Richon V, Sassone-Corsi P. 2013. Unraveling the secrets of the epigenome. *The Scientist* 27:62−63.

Uller T. 2013. Non-genetic inheritance and evolution. In: Kampourakis K (Ed.). *The Philosophy of Biology: A Companion for Educators.* Springer, New York, NY, pp. 267−287.

Chapter 23

1944
Genetic Material

THE STANDARD PARADIGM

Proteins (or perhaps some other complex class of organic macromolecule) must be the genetic material of life. During the first half of the 20th century most biochemists dismissed nucleic acids as being the stuff of heredity, because DNA molecules seemed far too structurally simple (being composed of only four distinct types of nucleotide subunits) to encode life's vast diversity and extraordinary complexity.

THE CONCEPTUAL REVOLUTION

DNA proved to be life's hereditary macromolecule, after all. Critical experiments demonstrating this surprising fact were conducted on bacteria by Oswald Avery and his colleagues in 1944, and by Alfred Hershey and Martha Chase on bacteriophages in 1952. With the benefit of hindsight, DNA as the carrier of hereditary information becomes intelligible, because − like the Morse code with its simple dots and dashes − even small numbers of different types of characters when strung together in extremely long chains can encode an indefinite variety of messages.

PS-score: 10

The discovery that nucleic acids (rather than proteins) are the genetic material of life surely merits a 10 on the paradigm-shift scale. Even so, an element of fortuity must be acknowledged regarding the broader significance of this revelation. Imagine (for the sake of argument) that the genetic material of at least some life forms other than bacteria and their phages eventually had proved to be proteins or some other biochemical substance; then the findings of Avery et al. and Hershey and Chase would not have been nearly so broadly impactful for the field of evolutionary genetics. Like many

J.C. Avise: Conceptual Breakthroughs in Evolutionary Genetics.
DOI: http://dx.doi.org/10.1016/B978-0-12-420166-8.00023-3

other basic discoveries in molecular genetics, universality or near-universality is key to the broader biological impact of a new discovery-based paradigm. In other words, we seldom tend to remember or praise "lesser" discoveries that may have been just as hard-won but that proved to be idiosyncratic to particular taxa. It is also worthwhile to remember just how recent (much less than a century ago) was the elucidation of DNA as the genetic material of life on Earth. This fact now seems so utterly basic and obvious.

REFERENCES AND FURTHER READING

Avery OT, MacLeod CM, McCarthy M. 1944. Studies on the chemical nature of the substance inducing transformation of pneumococcal types: induction of transformation by a deoxyribonucleic acid fraction isolated from *Pneumococcus* type III. *J. Exptl. Med.* 79:137–158.
Hershey AD, Chase M. 1952. Independent functions of viral protein and nucleic acid in growth of bacteriophage. *J. General. Physiol.* 36:39–56.

Chapter 24

1950
Jumping Genes

THE STANDARD PARADIGM

Each gene has a specific and stationary location along a chromosome.
From mapping experiments (see Chapter 14) in fruit flies and other organisms, scientists had mapped the presumably stable locations of many genes
in the karyotypes (chromosomal constitutions) of various eukaryotic species.

THE CONCEPTUAL REVOLUTION

Based on her work on maize conducted mostly during the 1940s, in 1950
Barbara McClintock shocked the scientific world by publicly announcing her
discovery of transposable elements. These "jumping genes" were documented
to shift positions (often replicatively) throughout the corn genome, with many
functional consequences for the host. We now know that transposable elements
comprise several categories of mobile DNA including various elements (such as
LINEs, SINEs, and LTRs that transpose and proliferate via an RNA intermediate that is reverse transcribed before reinsertion into the genome), and trans-
posons that shift their genomic positions by a cut-and-paste mechanism.

PS-score: 9

The discovery of transposable elements ranks among the most significant
evolutionary genetic discoveries in the latter half of the 20th century.
Following McClintock's lead, mobile elements of many types have been
found in astounding abundance in a wide variety of organisms. For example,
active or deceased mobile elements constitute at least 50% of the human
genome, and this figure is probably a severe underestimate due to the diffi-
culty of identifying elements that have partially decomposed since their time
of origin and spread. Transposable elements are also notable for their many
potential (and often realized) evolutionary consequences, such as being the

J.C. Avise: Conceptual Breakthroughs in Evolutionary Genetics.
DOI: http://dx.doi.org/10.1016/B978-0-12-420166-8.00024-5

sponsors of numerous *de novo* mutations and their documented proclivity to become coopted by host genomes into gene regulatory roles (see Chapter 52). Indeed, McClintock originally referred to jumping genes as both "mutable loci" and "controlling elements". Transposable elements also provided a primary stimulus for the important evolutionary concepts of selfish DNA (see Chapter 48) and parasitic DNA — the notion that quasi-autonomous snippets of genetic material can behave much like viral particles or other self-serving parasites. Some transposable elements are explicitly recognized as being at least quasi-parasitic because they are referred to as endogenous retroviruses (ERVs), which proliferate via reverse transcription (see Chapter 42) and collectively comprise about 8% of the human genome. For her discovery and early characterization of mobile elements, in 1983 McClintock became the only woman to receive an unshared Nobel Prize for Physiology or Medicine.

REFERENCES AND FURTHER READING

McClintock B. 1950. The origin and behavior of mutable loci in maize. *Proc. Natl Acad. Sci. USA* 36:344−355.

McClintock B. 1956. Controlling elements and the gene. *Cold Spring Harbor Symp. Quant. Biol.* 21:197−216.

McClintock B. 1956. Intranuclear systems controlling gene action and mutation. *Brookhaven Symp. Biol.* 16:13−47.

McDonald JF (Ed.). 1993. *Transposable Elements and Evolution*. Kluwer, Dordrecht, The Netherlands.

Saedler H, Gierl A. 1996. *Transposable Elements*. Springer, Berlin, Germany.

Chapter 25

1952
Pluripotency

THE STANDARD PARADIGM

During individual development, the genome of each differentiated somatic cell becomes irreversibly specialized. At that time it was universally believed that the genome of a liver cell (for example) could not direct the production of other types of cells or tissues. Cellular specialization (differentiation) during ontogeny presumably entails tissue-specific changes in genetic regulation (see Chapter 41) that convert away from what must have been the pluripotent condition of the zygote's genome and its resultant stem cells (which ultimately gave rise to the many different cell types in a growing individual).

THE CONCEPTUAL REVOLUTION

The new paradigm is that pluripotency must be retained by (or at least can be restored to) genomes taken even from highly differentiated somatic cells. This startling fact was revealed via nuclear transplantation (NT) experiments first conducted in the 1950s, when embryologists Robert Briggs and Thomas King microsurgically transferred cell nuclei from the embryos or tadpoles of frogs into artificially enucleated frog's eggs. These eggs then began to divide and multiply mitotically, eventuating in a new generation of tadpoles each of which was a clonal replica of its nuclear-donor parent. Evidently, genomes from differentiated cells of the donor could begin to act again like stem cells to direct full embryonic development under suitable circumstances. In 1962 John Gurdon and colleagues extended this approach using nuclei transplanted from intestinal skin cells of adult frogs, and in 1997 Ian Wilmut and his colleagues did much the same when they produced the famous lamb Dolly, the world's first NT-generated mammalian clone. Since then, artificial NT-cloning has been extended to a variety of domestic animal species, such as dogs, cows, and pigs. Although NT-cloning in humans proved to be

J.C. Avise: Conceptual Breakthroughs in Evolutionary Genetics.
DOI: http://dx.doi.org/10.1016/B978-0-12-420166-8.00025-7

considerably more difficult, the feat finally was accomplished in 2013 when Shoukhrat Mitalipov and his colleagues successfully fused enucleated human oocytes with skin cells from fetuses and babies. The hope is that the resulting blastocysts might soon become a reliable source of pluripotent embryonic stem cells with a wide range of potential applications in medicine. For his pioneering role in stem-cell research and NT-cloning, in 2012 John Gurdon received a Nobel Prize in Physiology or Medicine.

PS-score: 5

This paradigm shift gets only an intermediate score because its primary conceptual impact has been in the fields of stem-cell biology and development (rather than evolutionary genetics writ more broadly).

REFERENCES AND FURTHER READING

Briggs R, King TJ. 1952. Transplantation of living nuclei from blastula cells into enucleated frog's eggs. *Proc. Natl Acad. Sci. USA* 38:455–463.

Gurdon JB, Laskey RA, Reeves OR. 1975. The developmental capacity of nuclei transplanted from keratinized skin cells of adult frogs. *J. Embryol. Exp. Morphol.* 34:93–112.

Wilmut I, Schnieke AE, McWhir J, Kind AJ, Campbell KHS. 1997. Viable offspring derived from fetal and adult mammalian cells. *Nature* 385:810–813.

Avise JC. 2008. *Clonality: The Genetics, Ecology, and Evolution of Sexual Abstinence in Vertebrate Animals*. Oxford University Press, Oxford, UK.

Tachibana M, Amato P, Sparman M, Marti Gutierrez N, Tippner-Hedges R, Mitalipov S, (principal investigator), et al., 2013. Human embryonic stem cells derived by somatic cell nuclear transfer. *Cell* 153:1–11.

Vogel G. 2013. Human stem cells from cloning, finally. *Science* 340:795.

Chapter 26

1952
Aging

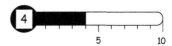

THE STANDARD PARADIGM

Aging (senescence) is hard to reconcile as an evolutionary outcome of natural selection. Senescence (defined as a persistent decline in the survival probability or reproductive output of an individual because of internal physiological deterioration) was a longstanding evolutionary enigma. Why would natural selection have permitted a state of affairs in which genetic predispositions for aging and death appear to be almost universal to multicellular life?

THE CONCEPTUAL REVOLUTION

In 1952, the Nobel Prize-winning immunologist Peter Medawar wrote an essay that laid the conceptual foundation for solving this puzzle. Medawar noted that natural selection inevitably declines in force through successive age cohorts in any age-structured population of self-reproducing entities. For example, the strength of natural selection on genes in 80-year-old humans is inevitably less than the strength of selection on the same genes in teenagers because teenagers' genes under any multi-generation scenario are destined on average to be survived by many more descendants than will those of 80-year-olds. Thus, senescence in any species is a feature that evolved as a logical consequence of the declining force of natural selection through successive age classes in a population. In effect, even a small advantage conferred early in life may outweigh a catastrophic disadvantage withheld until later, such that natural selection can be said to be biased in favor of youth whenever a conflict of interest arises.

Medawar's insight provided the ultimate evolutionary explanation for aging. It led in turn to several penultimate or more proximate evolutionary hypotheses regarding how diminishing selective pressures with age might translate into genetically hard-wired propensities for aging. For example, under the "mutation accumulation" hypothesis, later-age classes in effect become garbage bins where alleles with age-delayed deleterious somatic

J.C. Avise: Conceptual Breakthroughs in Evolutionary Genetics.
DOI: http://dx.doi.org/10.1016/B978-0-12-420166-8.00026-9

effects accumulate in evolution because of weak selection pressure there against their loss. By contrast, under the hypothesis of "antagonistic pleiotropy", alleles for aging are favored by natural selection because their beneficial effects at early stages of life outweigh any antagonistic deleterious effects later in life. An example might be a gene that promotes strong bones in youth but, as an ancillary byproduct, hardens arteries (promotes atherosclerosis) later in life.

PS-score: 4

Although Medawar's concepts paved the path to solving an evolutionary genetic mystery to nearly everyone's general satisfaction, I have given it only a modest *PS-score* because its direct implications and applications are confined mostly to the field of gerontology. Nevertheless, Medawar's reasoning also exemplifies evolutionary thought in the broader fields of life-history biology and demographic theory.

REFERENCES AND FURTHER READING

Medawar P. 1952. *An Unsolved Problem of Biology*. H.K. Lewis, London, UK.

Williams GC. 1957. Pleiotropy, natural selection, and the evolution of senescence. *Evolution* 11:398–411.

Bernstein C, Bernstein H. 1991. *Aging, Sex, and DNA Repair*. Academic Press, San Diego, CA.

Rose MR. 1991. *Evolutionary Biology of Aging*. Oxford University Press, New York, NY.

Roff D. 1992. *The Evolution of Life Histories: Theory and Analysis*. Chapman & Hall, New York, NY.

Stearns S. 1992. *The Evolution of Life Histories*. Oxford University Press, Oxford, UK.

Austad SN. 1997. *Why We Age: What Science is Discovering about the Body's Journey Through Life*. Wiley & Sons, New York, NY.

Chapter 27

1953
Origins of Life

THE STANDARD PARADIGM

The origin of life falls outside the emerging evolutionary genetic framework.
Before the macromolecules of heredity and biochemistry were well characterized (see Chapter 23), it seemed inconceivable to most scientists that life's
mode of origin could be understood in great detail, much less replicated in
the laboratory (see Chapter 2).

THE CONCEPTUAL REVOLUTION

Following earlier suggestions by Aleksandr Oparin (1924) and J.B.S. Haldane
(1929), in 1953 Stanley Miller and his doctoral advisor Harold Urey designed a
laboratory apparatus to simulate how organic molecules might have arisen from
inorganic compounds under primordial Earth conditions. By mixing ammonia,
methane, water, and hydrogen in a flask and subjecting this inorganic soup to
electrical discharges (simulating lightning), the researchers astonished the scientific world by showing that a wide variety of amino acids (the building blocks
of proteins) and other organic compounds can arise spontaneously under
suitable abiotic conditions. Later experiments by Fox and Dose (1977), among
others, went further by demonstrating that nucleotides (the building blocks of
nucleic acids) can arise in this general fashion too, and also that synthesized
amino acids sometimes link spontaneously into more complex organic
polymers.

PS-score: 3

The experiment by Miller was path-breaking because it opened everyone's
eyes to possible routes by which organic material could arise naturally from
inorganic non-life. On the other hand, this and many subsequent laboratory
trials have been disappointing in the sense that they have not yet managed to

J.C. Avise: Conceptual Breakthroughs in Evolutionary Genetics.
DOI: http://dx.doi.org/10.1016/B978-0-12-420166-8.00027-0

tweak the experimental conditions in such a way as to generate self-replicating genetic molecules that would enable natural selection to take full hold and thereby speed the evolutionary emergence of living organisms from inorganic substrates (see Chapter 55).

In an interesting footnote to this topic, in 2010 the genome pioneer Craig Venter held a press conference in which he announced the creation of "synthetic life". Venter's scientific team had artificially synthesized a large piece of DNA in the laboratory and inserted it into a bacterial cell, functionally taking over the latter. Critics have noted that this technological breakthrough is still a long way from creating living organisms from non-living starting points.

REFERENCES AND FURTHER READING

Oparin AI. 1924. *The Origin of Life*. Moskovsky Rabochiy, Moscow, Russia.

Haldane JBS. 1929. The origin of life. *Rationalist Annual* 1929:148–169.

Miller SL. 1953. A production of amino acids under possible primitive Earth conditions. *Science* 117:528–529.

Fox SW, Dose K. 1977. *Molecular Evolution and the Origin of Life*. Marcel Dekker, New York, NY.

Miller SL. 1992. The prebiotic synthesis of organic compounds as a step toward the origin of life. In: Schopf JW (Ed.). *Major Events in the History of Life*. Jones & Bartlett, Boston, MA, pp. 1–25.

Szostak JW, Bartel DP, Luisi PL. 2001. Synthesizing life. *Nature* 409:387–390.

Venter JC. 2011. Synthesizing life. *The Scientist* 25:60.

Chapter 28

1954
Hybridization

THE STANDARD PARADIGM

Novel mutations and genetic recombination introduce the genetic variation that provides the fodder for natural selection. The genetic variation that underlies population-level evolution traditionally was attributed to *de novo* mutations and − especially in sexual species − to genetic recombination during gametogenesis and syngamy (fertilization).

THE CONCEPTUAL REVOLUTION

According to Edgar Anderson and G. Ledyard Stebbins (1954), interspecific introgressive hybridization can be another wellspring of genetic variation upon which natural selection acts in some circumstances. Indeed, because hybridization in effect throws together whole suites of coadapted genes that formerly had evolved independently in separate species, introgression may have the potential to forge new adaptations more quickly and suddenly than by more conventional evolutionary routes. The article by Anderson and Stebbins was a paradigm buster not only for this reason but also because it contravened some other standard population genetic wisdoms at that time. Formerly, hybridization often had been cast in a negative light as a process that breaks up favorable gene combinations, or perhaps provides an evolutionary motivation for the erection of prezygotic isolating barriers between species to ameliorate what would otherwise be problems of hybrid sterility or infertility. Anderson and Stebbins' thesis generally cast a different conceptual aura on hybridization by suggesting that this widespread phenomenon can also serve as a positive evolutionary stimulus.

J.C. Avise: Conceptual Breakthroughs in Evolutionary Genetics.
DOI: http://dx.doi.org/10.1016/B978-0-12-420166-8.00028-2

PS-score: 3

Even if relatively common, introgression would not negate the pre-eminent roles for *de novo* mutation and standard genetic recombination during the evolutionary process. However, the Anderson–Stebbins argument certainly added an important new ingredient or flavor to many discussions about how evolution transpires. Furthermore, there has been a recent resurgence of interest in hybridization, with some authors (e.g., Arnold, 1997) arguing that introgression has been a major and oft-overlooked force in evolutionary genetics. Today, evolutionary geneticists utilize large numbers of nuclear and cytoplasmic molecular markers, from many linked and unlinked loci, to analyze hybrid zones and introgression patterns in exacting detail for a wide variety of animal and plant species.

REFERENCES AND FURTHER READING

Anderson E. 1949. *Introgressive Hybridization*. Wiley & Sons, New York, NY.

Anderson E, Stebbins GL. 1954. Hybridization as an evolutionary stimulus. *Evolution* 8:378–388.

Arnold ML. 1997. *Natural Hybridization and Evolution*. Oxford University Press, New York, NY.

Chapter 29

1954
Life's Antiquity

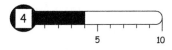

THE STANDARD PARADIGM

Life originated long after the Earth had formed. Darwin and other evolutionary biologists were well aware of fossils, but had few ways to date them beyond the relative gauge of stratigraphy itself — the realization that more recent forms of life typically appear above more ancient forms of life in stratified layers of sedimentary rock. Many paleontologists surmised that the origin of life (see Chapter 27) was so improbable an event that it must have taken many eons for life to emerge after the Earth's formation. Especially challenging to understand was the relatively sudden appearance in the fossil record of a wide variety of animal types at the Pre-Cambrian/Cambrian boundary, "merely" some 600 million years ago. This Cambrian explosion implied that plant lineages and microbes must have made their appearances much earlier, but exactly when remained a mystery.

THE CONCEPTUAL REVOLUTION

In 1954, the paleobotanist Elso Barghoorn and the geologist Stanley Tyler published a paper documenting fossil remains of plant microorganisms at least 2 billion years old. Subsequent fossil discoveries pushed the minimum age of life back even further, to at least 3.4 billion years ago. Thus, life apparently emerged on Earth far earlier than formerly had been supposed.

PS-score: 4

The great antiquity of life on Earth came as a surprise in the 1950s, but is taken for granted now. For his role in this conceptual conversion, and for his broader impact on the fields of organic chemistry, paleobiology, and the Earth sciences, Barghoorn is remembered today as "the father of Pre-Cambrian paleontology".

J.C. Avise: Conceptual Breakthroughs in Evolutionary Genetics.
DOI: http://dx.doi.org/10.1016/B978-0-12-420166-8.00029-4

REFERENCES AND FURTHER READING

Barghoorn ES, Tyler SA. 1954. Occurrence of structurally preserved plants in Pre-Cambrian rocks of the Canadian shield. *Science* 119:606–608.

Schopf JW (Ed.). 1983. *Earth's Earliest Biosphere, Its Origin and Evolution.* Princeton University Press, Princeton, NJ.

Gould SJ. 1989. *Wonderful Life: The Burgess Shale and the Nature of History.* Norton & Co, New York, NY.

Knoll AH. 2003. *Life on a Young Planet: The First Three Billion Years of Evolution on Earth.* Princeton University Press, Princeton, NJ.

Margulis L, Knoll AH. 2005. *Elso Sterrenberg Barghoorn, Jr, 1915–1984. Biogr. Memoirs* 87:3–11. The National Academies Press, Washington, DC.

Chapter 30

1956
Evolution in Action

THE STANDARD PARADIGM

Evolution by natural selection is far too slow a process to be observed directly.
Although Darwin had argued that responses to artificial selection (e.g., for various phenotypic features in dog breeds and domestic pigeons) could be quite rapid, the general sentiment remained that evolution by means of natural selection was normally much too sluggish for direct documentation during a human lifespan.

THE CONCEPTUAL REVOLUTION

In the mid-1950s, Bernard Kettlewell reported that the survival of peppered moths in England depended on how well the wing colors of these insects matched the tree trunks on which they rested. Kettlewell's empirical evidence, which came from experiments and observations made directly on field plots, showed that predation (and crypsis against it) was potent enough to cause measurable evolutionary-genetic responses in moth populations across even one or a few insect generations.

PS-score: 5

Kettlewell's work was path-breaking, but later many more examples of evolution in action were well documented by experimentalists and astute field biologists. Indeed, career-long research programs have been built around direct observations of evolutionary changes in creatures ranging from European snails and Galapagos finches in the field, to fruit flies in population cages, to bacterial populations in laboratory chemostats. For example, the husband–wife team of Peter and Rosemary Grant have devoted their careers to studying the evolutionary dynamics of changes in bill sizes (often as a function of selection pressures from temporally variable food conditions) of seed-eating landbirds

J.C. Avise: Conceptual Breakthroughs in Evolutionary Genetics.
DOI: http://dx.doi.org/10.1016/B978-0-12-420166-8.00030-0
63

on the Galapagos Islands (which Charles Darwin himself had found to be such a fascinating evolutionary theater a century earlier).

REFERENCES AND FURTHER READING

Kettlewell HBD. 1956. Further selection experiments on industrial melanism in the Lepidoptera. *Heredity* 10:287–301.

Ford EB. 1964. *Ecological Genetics*. Chapman & Hall, New York, NY.

Kettlewell HBD. 1973. *The Evolution of Melanism: The Study of a Recurring Necessity*. Clarendon Press, Oxford, UK.

Ford EB. 1981. *Taking Genetics into the Countryside*. Weidenfeld & Nicholson, London, UK.

Endler JA. 1986. *Natural Selection in the Wild*. Princeton University Press, Princeton, NJ.

Hutchinson EW, Shaw AJ, Rose MR. 1991. Quantitative genetics of postponed aging in *Drosophila melanogster*. II. Analysis of selected lines. *Genetics* 127:729–737.

Lenski RE, Travisiano M. 1994. Dynamics of adaptation and diversification: a 10,000-generation experiment with bacterial populations. *Proc. Natl Acad. Sci. USA* 91:6808–6814.

Weiner J. 1994. *The Beak of the Finch*. Cambridge University Press, Cambridge, UK.

Palumbi SR. 2001. *The Evolution Explosion: How Humans Cause Rapid Evolutionary Change*. Norton & Co., New York, NY.

Grant PR, Grant BR. 2008. *How and Why Species Multiply*. Princeton University Press, Princeton, NJ.

Grant PR, Grant BR (Eds). 2010. *In Search of the Causes of Evolution. From Field Observations to Mechanisms*. Princeton University Press, Princeton, NJ.

The 1960s and 1970s: Dawn of the Molecular Era

The conceptual breakthroughs described in Part III reside primarily in the early molecular era of evolutionary genetics, after it was discovered that natural populations of most organisms harbor great stores of genetic variation in proteins. They begin in 1963 with the then-radical notion that biological macromolecules can be a rich source of phylogenetic information, and they conclude in 1979 with several breakthroughs, including the birth of phylogeography and the death of the concept that the genetic code is universal. Among the many other notable achievements during the 1960s and 1970s were the invention and elaboration of the diverse concepts of kin selection, coevolutionary interactions, molecular clocks, cladistics, levels of selection, endosymbiosis, genetic regulation, neutrality theory, sperm competition, cryptic female choice, selfish genes, split genes, and the refurbished idea of exaptations.

Chapter 31

1963 Molecular Phylogeny

THE STANDARD PARADIGM

Organismal phenotypes provide the characters for phylogeny reconstruction. In the first 100 years following Darwin, scientists estimated phylogenetic trees for various taxa by comparing visible organismal phenotypes — e.g., morphological, physiological, or behavioral characteristics — that they could only presume must reflect the true evolutionary genetic relationships of organisms.

THE CONCEPTUAL REVOLUTION

In the 1950s and 1960s, several molecular technologies were developed that eventually would give direct and quick access to voluminous genetic information recorded in DNA and protein sequences. In 1963, Emanuel Margoliash introduced the revolutionary idea that organismal relationships could be deduced from such macromolecular data. By compiling published information on the molecular strings (each 104 amino acids long) of the cytochrome *c* protein from several sources — human, pig, horse, rabbit, chicken, tuna, and yeast — Margoliash concluded, "the extent of variation among cytochromes *c* is compatible with the known phylogenetic relations of species. Relatively closely related species show few differences . . . phylogenetically distant species exhibit wider dissimilarities." In 1967, Fitch and Margoliash pioneered a formal phylogenetic procedure for constructing evolutionary trees from such molecular sequences. More recently, molecular phylogenetics has become a flourishing enterprise with many available algorithms suited for different categories of genetic data.

J.C. Avise: Conceptual Breakthroughs in Evolutionary Genetics.
DOI: http://dx.doi.org/10.1016/B978-0-12-420166-8.00031-2

PS-score: 8

Organismal-level phenotypes continue to be a useful and ready source of phylogenetic information from many taxa, but generally they have been superseded by molecular data for detailed phylogenetic appraisals. Molecular characters are of special significance because they are: (1) unambiguously genetic; (2) ubiquitously distributed across all forms of life (including microbial); (3) quantifiable and copious in number; (4) arguably less prone to homoplasy (evolutionary convergences or reversals) that otherwise can complicate phylogeny estimation from adaptive phenotypic traits; and (5) arguably less variable in their evolutionary rates (see Chapter 34). Today, molecular phylogenetics is a robust and multifaceted discipline, with molecular phylogenies routinely being published for a wide diversity of taxa. Indeed, it is not hard to imagine a time in the not-too-distant future when the success of molecular phylogenetics leads to the field's inevitable demise, because molecular phylogeneticists will have reconstructed (about as well as is theoretically possible) essentially all major branches and twigs in the Earth's tree of life. Thus, this successful enterprise will have put itself out of business.

REFERENCES AND FURTHER READING

Margoliash E. 1963. Primary structure and evolution of cytochrome *c*. *Proc. Natl Acad. Sci. USA* 50:672–679.

Fitch WM, Margoliash E. 1967. Construction of phylogenetic trees. *Science* 155:279–284.

Hillis DM, Moritz C, Mable BK. 1996. *Molecular Systematics*, 2nd edition. Sinauer, Sunderland, MA.

Page RDM, Holmes EC. 1998. *Molecular Evolution: A Phylogenetic Approach*. Blackwell, Oxford, UK.

Nei M, Kumar S. 2000. *Molecular Evolution and Phylogenetics*. Oxford University Press, New York, NY.

Felsenstein J. 2004. *Inferring Phylogenies*. Sinauer, Sunderland, MA.

Chapter 32

1964
Kin Selection

THE STANDARD PARADIGM

Natural selection operates via the differential reproductive successes of con-specific organisms (i.e., via differences in personal genetic fitness).

THE CONCEPTUAL REVOLUTION

In 1964, the British evolutionary theorist William Hamilton proposed that the concept of reproductive fitness should incorporate not only each individual's personal reproduction but also the probability that a focal individual's genes are successfully transmitted through genetic relatives. These latter transmission probabilities are influenced by the coefficients of relatedness (proportions of genes shared) by the relevant kin. Hamilton thereby introduced the notion of "inclusive fitness", which encompasses an individual's personal fitness and the fitness of his or her relatives. His insight raised the then-radical notion that a gene could be evolutionarily favored even if it tends to diminish an individual's personal reproduction, provided that more than compensatory numbers of the gene's copies are transmitted through the focal individual's kin. In other words, an individual's genes can profit under the mathematics of inclusive fitness by being transmitted through biological relatives as well as through the individual itself.

PS-score: 7

Hamilton had laid out a revolutionary new framework for viewing selection and altruism, topics that are central to studies of the biology of social interactions. His papers introduced the concept of "kin selection", which became widely viewed as an essential consideration in much of sociobiology. The idea that kin selection underlies altruism led to Hamilton's rule: an allele encoding a social behavior will tend to increase in frequency in a population if r (the coefficient

J.C. Avise: Conceptual Breakthroughs in Evolutionary Genetics.
DOI: http://dx.doi.org/10.1016/B978-0-12-420166-8.00032-4
69

of genetic relatedness between two actors) is greater than the ratio of the cost (C) of the behavior (loss in personal fitness through self-sacrifice) compared to the benefit (B) received via increased reproduction by relatives ($r > C/B$). Hamilton's rule has been invoked to help explain many seemingly altruistic phenomena in nature, ranging from apparent self-sacrifice in humans to the highly eusocial behaviors of ants and bees. On the other hand, the importance of kin selection in evolutionary genetics has also come under criticism and remains subject to ongoing debate (see, for example, Wilson and Wilson, 2007; Nowak et al., 2010).

REFERENCES AND FURTHER READING

Hamilton WD. 1964. The genetical evolution of social behavior I. *J. Theoret. Biol.* 7:1−16.

Hamilton WD. 1964. The genetical evolution of social behavior II. *J. Theoret. Biol.* 7:17−52.

Wilson EO. 1975. *Sociobiology*. Belknap Press, Cambridge, MA.

Crozier RH, Pamilo P. 1996. *Evolution of Social Insect Colonies: Sex Allocation and Kin Selection*. Oxford University Press, Oxford, UK.

Hamilton WD. 1996. *Narrow Roads of Gene Land, I. Evolution of Social Behavior*. Oxford University Press, Oxford, UK.

Wilson DS. 2002. *Darwin's Cathedral: Evolution, Religion, and the Nature of Society*. University of Chicago Press, Chicago, IL.

Wilson DS, Wilson EO. 2007. Rethinking the theoretical foundations of sociobiology. *Quart. Rev. Biol.* 82:327−348.

Nowak M, Tarnita CE, Wilson EO. 2010. The evolution of eusociality. *Nature* 466:1057−1062.

Wilson EO. 2012. *The Social Conquest of Earth*. Norton & Co., London, UK.

Segerstrale U. 2013. *Nature's Oracle: A Life of W.D. Hamilton*. Oxford University Press, Oxford, UK.

Chapter 33

1964
Coevolution

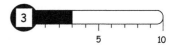

THE STANDARD PARADIGM

Evolution takes place as each species gradually adapts to its physical environment. Although Darwin and many other evolutionary biologists were certainly attuned to species' interactions as being an important component of the selective milieu for each species, it remained easy to overlook some kinds of biotic interactions because no single general term existed to describe such interspecific selective interplay.

THE CONCEPTUAL REVOLUTION

In 1964, ecologists Paul Ehrlich and Peter Raven introduced the term "coevolution". In that classic paper they focused on interactions between plants and phytophagous insects, emphasizing the intimacy and consequentiality of the reciprocal coevolutionary dances between these two sets of participants. Today, coevolution is defined as the joint evolution of two or more ecologically interacting species, each of which evolves in response to selective pressures imposed by the other. Predator−prey, host−parasite, pollinator−host, and various symbioses are among the many examples in which species that interact in ecological time impose selective pressures that mutually impact one another across evolutionary time.

PS-score: 3

Sometimes a novel word or phrase encapsulates a concept and becomes part of the working lexicon of a discipline. Examples of evocative yet utilitarian terms that helped to define and stimulate emerging fields of biological inquiry include kin selection and sociobiology (see Chapter 32), sperm competition (Chapter 43), and phylogeography (Chapter 51), among many others. Actually, "coevolution" was defined only loosely by Ehrlich and Raven − a fact that

J.C. Avise: Conceptual Breakthroughs in Evolutionary Genetics.
DOI: http://dx.doi.org/10.1016/B978-0-12-420166-8.00033-6

allowed for a variety of different interpretations by subsequent authors. For example, coevolution has been used to describe biotic interactions ranging from intimate connections between pairs of intracellular symbionts (see Chapter 38) to manifold and often diffuse species' interactions that collectively shape the structure of entire natural communities or ecosystems. Despite this definitional vagueness, the broader message from Ehrlich and Raven's seminal paper still rings loud. Coevolution is a key point of contact between evolution and ecology, in the sense that organisms themselves must be considered an important part of the environment that exerts selective pressures upon all species.

REFERENCES AND FURTHER READING

Ehrlich PR, Raven PH. 1964. Butterflies and plants: a study in coevolution. *Evolution* 18:586–608.

Thompson JN. 1982. *Interaction and Coevolution.* Wiley, New York, NY.

Futuyma DJ, Slatkin M. (Eds). 1983. *Coevolution.* Sinauer, Sunderland, MA.

Thompson JN. 1994. *The Coevolutionary Process.* University of Chicago Press, Chicago, IL.

Chapter 34

1965
Molecular Clocks

THE STANDARD PARADIGM

Biological macromolecules, like organismal phenotypes, evolve at highly vari-
able tempos and modes within and across organismal lineages. In other words,
molecular evolution was thought to be an erratic or idiosyncratic process.

THE CONCEPTUAL REVOLUTION

The new paradigm was that at least some molecules evolve with sufficient
regularity as to provide molecular timepieces. Emile Zuckerkandl and Linus
Pauling were the first to propose the concept of a "molecular clock". The idea
fits well with the neutrality theory of molecular evolution (see Chapter 40)
because the rate of neutral evolution in DNA sequences is in principle equal
to the mutation rate to neutral alleles. However, molecular clocks need not be
incompatible with natural selection, because if large numbers of DNA
sequences are acted upon by multifarious selection pressures over long periods
of evolutionary time, then short-term fluctuations in selection could cancel out
such that genetic distances between taxa might correlate well with times
elapsed since common ancestry.

Two aspects of the molecular clock concept should not be misconstrued.
First, the proposal is not that the clock is metronomic, but rather that it might
tick at a stochastically constant rate, like radioactive decay. Second, molecular
clocks may be helpful but they are not prerequisite for the use of DNA
sequences in molecular systematics (see Chapter 31), either because many tree-
building algorithms relax assumptions of rate homogeneity among genes and
lineages, or because they deal with raw sequence characters before the latter are
converted into estimates of inter-taxon genetic distance (see Chapter 35).

J.C. Avise: Conceptual Breakthroughs in Evolutionary Genetics.
DOI: http://dx.doi.org/10.1016/B978-0-12-420166-8.00034-8

PS-score: 6

In countless empirical studies conducted during the past half-century, various DNA sequences have proved to evolve at heterogeneous rates at several levels: across nucleotide positions within a codon; among non-homologous genes within a lineage; among different classes of DNA within a genome; and among different genomes within an organismal lineage. Thus, there is no single molecular clock but rather a large ensemble of clocks ticking at different rates. This may seem disconcerting, but it also means that researchers can choose molecular clocks that are tailored to the approximate evolutionary timescale of each exercise in phylogenetic dating. For example, ribosomal RNA sequences have been extremely informative in reconstructing ancient events in the history of life (see Chapter 50), whereas rapidly evolving mitochondrial DNA sequences have revolutionized phylogenetic appraisals at the far more recent timescales of intraspecific differentiation (see Chapter 51).

REFERENCES AND FURTHER READING

Zuckerkandl E, Pauling L. 1965. Evolutionary divergence and convergence in proteins. In: Bryson V, Vogel HJ. (Eds). *Evolving Genes and Proteins*. Academic Press, New York, NY, pp. 97–166.

Ayala FJ. 1985. On the virtues and pitfalls of the molecular evolutionary clock. *J. Heredity* 77:226–235.

Nei M. 1987. *Molecular Evolutionary Genetics*. Columbia University Press, New York, NY.

Li W-H. 1997. *Molecular Evolution*. Sinauer, Sunderland, MA.

Hedges SB, Kumar S. (Eds). 2009. *The Timetree of Life*. Oxford University Press, Oxford, UK.

Chapter 35

1966
Cladistics

THE STANDARD PARADIGM

Overall organismal resemblance is the basis for biological classification and phylogeny reconstruction. Until the mid-1900s, following the tradition of Carolus Linnaeus in the 1700s, scientists generally classified organisms by gestalt appraisals of their broad phenotypic resemblance to one another. This "phenetic" approach, which predated the revolution in molecular systematics (see Chapters 31 and 34) reached its apogee in 1963 with the publication of *Principles of Numerical Taxonomy* by Robert Sokal and Peter H.A. Sneath, who argued that organisms should be grouped according to their overall similarity as quantified by defined rules using as many traits as possible. Philosophically, numerical taxonomists aimed to develop classification methods that were objective, explicit, and repeatable.

THE CONCEPTUAL REVOLUTION

The replacement worldview is that phylogeny should be appraised (and classifications based) not on the overall similarity between organisms, but rather on a specific subset of homology attributable to synapomorphic or shared-derived characters. This new phylogenetic paradigm became extraordinarily popular following the 1966 translation from German into English of Willi Hennig's *Phylogenetic Systematics*. The conceptual revolution that Hennig started became known as cladistics or cladism, because it seeks to identify clades (monophyletic units) in an evolutionary tree and in its associated classification. Cladists typically focus on reconstructing the branching component (cladogenetic aspect) of evolutionary trees, rather than on branch lengths or anagenesis. The ultimate goal is to develop organismal classifications based on correctly inferred cladogenetic patterns.

J.C. Avise: Conceptual Breakthroughs in Evolutionary Genetics.
DOI: http://dx.doi.org/10.1016/B978-0-12-420166-8.00035-X

In the 1960s and 1970s, extensive philosophical debates took place between the pheneticists (numerical taxonomists) and the cladists about the optimal strategies for evolutionary reconstructions and classification schemes.

PS-score: 7

The philosophical pillar of cladism — that genetic synapomorphies (molecular or otherwise) properly earmark clades — was a major conceptual breakthrough that is now acknowledged almost universally. It distinguished similarity indicative of shared ancestry from similarity due either to convergent evolution (homoplasy) or to the shared retention of ancestral traits (sympleisomorphy). In addition to clarifying the concept of homology and the origins of organismal resemblance, cladism transformed the fields of phylogenetics and systematics and provided the intellectual basis for parsimony methods that would come to dominate evolutionary reasoning in the ensuing decades.

REFERENCES AND FURTHER READING

Sokal RR, Sneath PHA. 1963. *Principles of Numerical Taxonomy*. Freeman & Co., San Francisco, CA.

Hennig W. 1966. *Phylogenetic Systematics*. University of Illinois Press, Urbana, IL [originally published in German in 1950].

Felsenstein J. 1983. Parsimony in systematics: biological and statistical issues. *Annu. Rev. Ecol. Syst.* 14:313–333.

Felsenstein J. 2004. *Inferring Phylogenies*. Sinauer, Sunderland, MA.

Wheeler WC. 2012. *Systematics*. Wiley-Blackwell, New York, NY.

Schmitt M. 2013. *From Taxonomy to Phylogenetics: Life and Work of Willi Hennig*. Brill, Leiden, The Netherlands.

Chapter 36

1966
Individual Selection

THE STANDARD PARADIGM

Although natural selection is a pervasive evolutionary force, the level at which it operates to greatest effect remained debatable. In particular, some evolutionary biologists still invoked group selection to account for a wide variety of biological outcomes. For example, a group selectionist argument might be that a population's sex ratio is nearly 1 : 1 so as to minimize conflicts over mate acquisition (see Chapter 18), or that organisms senesce and die (see Chapter 26) so as to make room for the next generation. Such group-selectionist perspectives on evolution reached their strongest advocacy in a 1962 book by Vero Wynne-Edwards, who argued (we would now say fallaciously) that natural selection routinely acts directly for the good of a group, a population, or a species.

THE CONCEPTUAL REVOLUTION

In 1966, George Williams published a devastating critique of group selectionism, arguing instead that natural selection operates with greatest force on inter-individual genetic variation in reproductive success (i.e., in genetic fitness).

PS-score: 7

Although Williams' advocacy for individual-level selection was not entirely novel then or now (the idea likewise has been championed by many other prominent biologists, including the ethologist David Lack and the evolutionary geneticists Robert Trivers and John Maynard-Smith), it nonetheless spearheaded a revolution in evolutionary thought with ripple effects that continue to reverberate to the present. With few exceptions, the kinds of naïve group selectionist formulations that formerly were in vogue are now generally viewed as being hopelessly inadequate if not flatly incorrect (but see Chapter 32, on kin selection). Instead, in most cases natural selection can best be interpreted as

J.C. Avise: Conceptual Breakthroughs in Evolutionary Genetics.
DOI: http://dx.doi.org/10.1016/B978-0-12-420166-8.00036-1

acting most efficaciously on genetic variation in reproductive fitness among individuals and families (or perhaps on their constituent genes; see Chapter 48).

REFERENCES AND FURTHER READING

Wynne-Edwards VC. 1962. *Animal Dispersion in Relation to Social Behavior*. Oliver & Boyd, Edinburgh, UK.

Williams GC. 1966. *Adaptation and Natural Selection*. Princeton University Press, Princeton, NJ.

Wilson DS. 1980. *The Natural Selection of Populations and Communities*. Benjamin-Cummings, Menlo Park, CA.

Williams GC. 1992. *Natural Selection: Domains, Levels, and Challenges*. Oxford University Press, Oxford, UK.

Alcock J. 2001. *The Triumph of Sociobiology*. Oxford University Press, Oxford, UK.

Parker GA. 2006. Behavioural ecology: The science of natural history. In: Lucas JR, Simmonos LW. (Eds). *Essays on Animal Behavior: Celebrating 50 years of Animal Behaviour*. Elsevier, Burlington, MA, pp. 23–56.

Chapter 37

1966
Genetic Variation

THE STANDARD PARADIGM

The magnitude of genetic variation in natural populations is open for debate.
Prior to the molecular era that began in the mid-1960s, the magnitude of
genetic variation in natural populations remained controversial, with many evolutionary biologists falling into either of two opposing camps: the classicists
who maintained that genetic variability in most species was low, such that conspecifics were homozygous for the same "wild-type" allele at most genetic
loci; and proponents of the balance view who argued that genetic variation was
extensive such that most loci were polymorphic and individuals typically were
heterozygous at a substantial fraction of their genes. A central tenet of the classical school was the concept of genetic load (see Chapter 17): the idea that
genetic variation in a population produces a heavy burden of reduced genetic
fitness. In support of the classical view was the observation that many *de novo*
mutations are deleterious. In contrast, the balance school tended to view
genetic polymorphisms as beneficial and maintained in populations by one or
another form of balancing natural selection (such as environmental heterogeneity, heterosis, or frequency-dependent fitness advantage). Two empirical
observations sometimes interpreted in favor of the balance view were as follows: phenotypic variation is often extensive in natural populations for many
traits; and artificial (human-mediated) selection for various phenotypic attributes was usually successful in many species, thus suggesting that genetic variation must underlie such traits. Nevertheless, by hard criteria such observations
were inconclusive for firmly resolving the classical/balance debate. Biologists
needed direct information on genetic variation from molecular biology.

THE CONCEPTUAL REVOLUTION

Empirically, population genetic variation proved to be unambiguously high. In
1966, three independent research laboratories published the first estimates of

J.C. Avise: Conceptual Breakthroughs in Evolutionary Genetics.
DOI: http://dx.doi.org/10.1016/B978-0-12-420166-8.00037-3

genetic variability based on a newly introduced molecular technique: multi-locus protein electrophoresis. The empirical results from these "allozyme" ana-lyses were clear: in organisms as different as humans and fruit flies, individual heterozygosities (the fractions of heterozygous loci) were substantial (often >10%) and a majority of assayed genes displayed molecular polymorphisms.

These findings were so unambiguous that one might suppose they settled the debate soundly in favor of the balance school. However, in an interesting turn of events, the classicists in effect regrouped into what became the neo-classical school of thought or the neutralists (see Chapter 40). The neutralists could not (and did not) deny that molecular genetic variation in most natural populations is extensive, but rather they questioned the degree to which balanc-ing natural selection played an active role in its maintenance. According to the neutralists, most genetic polymorphisms at the molecular level are selectively fitness-neutral.

PS-score: 7

This breakthrough gets a rather high score for two reasons: overnight, it trans-formed evolutionary thought about the extensive nature of molecular-level genetic variation; and the protein-electrophoretic technique went on to become the first of many molecular methods that collectively have revolutionized how genetic variation is tapped to provide molecular markers for a wide range of research applications in population ecology, ethology, and evolution.

REFERENCES AND FURTHER READING

Harris H. 1966. Enzyme polymorphism in man. *Proc. R. Soc. Lond. B* 164:298–310.

Johnson FM, Kanapi CG, Richardson RH, Wheeler MR, Stone WS. 1966. An analysis of poly-morphisms among isozyme loci in dark and light *Drosophila ananassae* strains from America and Western Samoa. *Proc. Natl Acad. Sci. USA* 56:119–125.

Lewontin RC, Hubby JL. 1966. A molecular approach to the study of genic heterozygosity in natural populations. II. Amount of variation and degree of heterozygosity in natural popula-tions of *Drosophila pseudoobscura. Genetics* 54:595–609.

Wallace B. 1970. *Genetic Load.* Prentice-Hall, Englewood Cliffs, NJ.

Lewontin RC. 1974. *The Genetic Basis of Evolutionary Change.* Columbia University Press, New York, NY.

Avise JC. 1994. *Molecular Markers, Natural History, and Evolution.* Chapman & Hall, New York, NY [a revised 2nd edition was published in 2004 by Sinauer, Sunderland, MA].

Chapter 38

1967
Organelle Origins

THE STANDARD PARADIGM

Constituent parts of a eukaryotic cell, including its internal organelles ("miniature organs"), evolved via conventional evolutionary processes. In 1890, the German histologist Richard Altmann first reported the existence within cells of "bioblasts" that later (in 1898) would be named mitochondria. Altmann suggested that these tiny intracellular structures carry out processes vital to their hosts and, remarkably, he even noted that they behave quite like elementary organisms. Still, the general sentiment remained that such organelles were standard cytoplasmic structures that had arisen and evolved *in situ* within eukaryotic cells.

THE CONCEPTUAL REVOLUTION

Scientists had long suspected that particular organelles (notably mitochondria and chloroplasts) contain their own tiny snippets of genetic material, separate from the cell's primary genome housed within its nucleus (see Chapter 13). In 1967, Lynn Margulis (writing at that time under the name Lynn Sagan) put forward a bold hypothesis that generally echoed Altmann's sentiment by proposing that these DNA-containing organelles originated in the distant past via symbiotic mergers when free-living bacteria invaded or were engulfed by proto-eukaryotic host cells. In other words, the evolutionary precursors of mitochondria and chloroplasts literally had been elementary autonomous organisms prior to these endosymbiotic unions. This panoramic reinterpretation of the chimeric etiology of eukaryotic cells soon became known as the endosymbiont theory of organelle origins.

J.C. Avise: Conceptual Breakthroughs in Evolutionary Genetics.
DOI: http://dx.doi.org/10.1016/B978-0-12-420166-8.00038-5

PS-score: 8

Margulis' scenario ranks high on my list of paradigm busters for two main reasons. First, over the past half-century her then-radical idea has withstood extensive scrutiny and debate, to the extent that the scientific community now accepts it almost universally. For example, one convincing line of argumentation stems from the molecular observation that small-subunit sequences of organellar ribosomal RNAs are phylogenetically closer to those of bacteria than they are to those of their eukaryotic hosts (see Chapter 50). Today, almost all biologists agree that the evolutionary precursors of mitochondria (the energy-generating powerhouses of cells) and chloroplasts (which play key photosynthetic roles in plants) probably did indeed exist as free-living microbes prior to their endosymbiotic incorporations into eukaryotic cells. Second, the endosymbiont theory was revisionary in the sense that it also contributed to the oft-overlooked notion that cooperation and symbiosis (as opposed to Darwinian conflict and strife) have been pervasive shaping forces during the evolutionary process.

REFERENCES AND FURTHER READING

Sagan L. 1967. On the origin of mitosing cells. *J. Theoret. Biol.* 14:225–274.

Margulis L. 1970. *Origin of Eukaryotic Cells*. Yale University Press, New Haven, CT.

Strassmann JE, Queller DC, Avise JC, Ayala FJ. 2011. *In the Light of Evolution, V: Cooperation and Conflict*. National Academies Press, Washington, DC.

Chapter 39

1968
Genomic Structure

THE STANDARD PARADIGM

Most genes come in a single copy per haploid genome. The standard sentiment at that time was that genes are like individual beads successively strung along a chromosomal string (see Chapters 9 and 14). In other words, each gene was represented as a single copy per haploid eukaryotic genome.

THE CONCEPTUAL REVOLUTION

During the 1950s and 1960s, a paradigm shift took place as biologists gradually accumulated evidence that much of the genome in any eukaryotic species consists of repetitive elements. No single discovery triggered this conceptual revolution, but a seminal paper by the geneticists Roy Britten and David Kohne in 1968 helped to crystallize and popularize this shifting attitude about the fundamental structure of eukaryotic genomes. Some of the earliest evidence for repetitive DNA came from DNA–DNA hybridization studies of various species in which it was noted that artificially dissociated strands of DNA reannealed in a test tube much faster than would be expected if each gene existed in only a single copy per genome. The reassociation kinetics instead implied that many DNA sequences in a genome must be moderately or highly repetitive.

PS-score: 8

Today, many categories of repetitive sequences are recognized, including simple or low-copy gene duplications, microsatellites (extensive arrays of individually very short tandem repeats), minisatellites (tandem arrays of somewhat longer repeats), and various categories of mobile elements that may be dispersed throughout the genome (see Chapter 24). Speculation about the functional significance of repetitive DNAs is equally diverse. On the

J.C. Avise: Conceptual Breakthroughs in Evolutionary Genetics.
DOI: http://dx.doi.org/10.1016/B978-0-12-420166-8.00039-7

positive side of the ledger, one common conceptual theme is that repetitiveness confers genetic redundancy and thereby facilitates the evolution of new operational capabilities in the host organism. Such enhanced cellular capacities might include the production in abundance of some much-needed protein or RNA product, novel or better opportunities to properly regulate a cell's gene expression, architectural roles in chromatin structure, and many others. Another distinct possibility is that some classes of repetitive elements (notably jumping genes) are mostly genomic garbage, perhaps often serving no functional role beyond their own selfish proliferation (see Chapter 48). In truth, these two viewpoints on repetitive DNA (valuable resource versus trash) are not necessarily in evolutionary opposition. Even if many repetitive sequences proliferated as selfish or parasitic elements, they may often become evolutionarily coopted by the host genome to perform useful cellular functions (see Chapter 52).

REFERENCES AND FURTHER READING

Britten RJ, Kohne DE. 1968. Repeated sequences in DNA. *Science* 161:529–540.

Ohno S. 1970. *Evolution by Gene Duplication.* Springer-Verlag, New York, NY.

Britten RJ, Davidson EH. 1971. Repetitive and non-repetitive DNA sequences and a speculation on the origins of evolutionary novelty. *Q. Rev. Biol.* 46:111–138.

Goldstein DB, Schlötterer C. 1999. *Microsatellites: Evolution and Applications.* Oxford University Press, Oxford, UK.

Shapiro JA, von Sternberg R. 2005. Why repetitive DNA is essential to genome function. *Biol. Rev. Camb. Philos. Soc.* 80:227–250.

Lynch M. 2007. *The Origins of Genome Architecture.* Sinauer, Sunderland, MA.

Fontdevila A. 2011. *The Dynamic Genome: A Darwinian Approach.* Oxford University Press, Oxford, UK.

Chapter 40

1968 Neutrality Theory

THE STANDARD PARADIGM

Biological macromolecules, like organismal phenotypes, evolve primarily by natural selection. There was no particular reason to believe that genetic macromolecules (nucleic acids and proteins) play by different evolutionary ground rules than do phenotypic traits. In other words, natural selection is probably pervasive at the molecular level.

THE CONCEPTUAL REVOLUTION

The new argument was that most genetic variation at the molecular level is neutral or nearly neutral with respect to genetic fitness. As stated by Kimura (1991), "the great majority of evolutionary mutant substitutions at the molecular level are caused by random fixation, through sampling drift, of selectively neutral (i.e., selectively equivalent) mutants under continued mutation pressure." This sentiment is known as the neutral theory. Neutrality concepts were introduced in the late 1960s by Motoo Kimura, and they immediately garnered widespread attention due in part to a paper by Jack King and Thomas Jukes provocatively titled "Non-Darwinian evolution . . . ".

The neutrality school of thought had strong roots in the quantitative tradition of theoretical population genetics developed much earlier in the 20th century (see Chapter 19). It predicts the level of genetic variation within a population and the rate of molecular evolution as a function of mutation rate, gene flow (where applicable), and population size, but conspicuously absent from the calculations are selection coefficients (because the relevant alleles are assumed to be neutral). Two misconceptions must be avoided. Neutrality theory neither suggests that most alleles are dispensable (of course they are not), nor does it deny that many *de novo* mutations are eliminated by natural selection because they are deleterious. Instead, the focus of neutrality theory is on the many alleles presumed to be involved in fitness-neutral genetic polymorphisms.

J.C. Avise: Conceptual Breakthroughs in Evolutionary Genetics.
DOI: http://dx.doi.org/10.1016/B978-0-12-420166-8.00040-3

PS-score: 8

Remarkably, within a decade of its initial formulation (and continuing today), neutrality theory gained sufficient acceptance to become molecular evolution's gigantic null hypothesis — the simplest conceptual framework for interpreting molecular evolution and the theoretical construct whose predictions must be falsified before alternative proposals invoking natural selection are to be seriously entertained. Neutrality theory is now nearly a half-century old, yet it continues to be a primary touchstone for nearly all basic research in molecular phylogenetics and molecular evolution.

REFERENCES AND FURTHER READING

Kimura M. 1968. Evolutionary rate at the molecular level. *Nature* 217:624–626.

Kimura M. 1968. Genetic variability maintained in a finite population due to mutational production of neutral and nearly neutral isoalleles. *Genet. Res.* 11:247–269.

King JL, Jukes TH. 1969. Non-Darwinian evolution: random fixation of selectively neutral mutations. *Science* 164:788–798.

Kimura M. 1983. *The Neutral Theory of Molecular Evolution*. Cambridge University Press, Cambridge, UK.

Kimura M. 1991. Recent developments of the neutral theory viewed from the Wrightian tradition of theoretical population genetics. *Proc. Natl Acad. Sci.* 88:5969–5973.

Chapter 41

1969
Gene Regulation

THE STANDARD PARADIGM

Sequence changes in structural genes (those that encode functional proteins) are responsible for most phenotypic evolution in multicellular organisms. In other words, conventional thought was that protein evolution was at the heart of organismal evolution. Although genetic mechanisms for the regulation of protein-coding genes in bacteria had been identified by Francois Jacob and Jacques Monod in 1961 (a feat for which they later shared a Nobel Prize), the notion that gene regulation might be broadly important in the evolution of multicellular life had not yet percolated widely in the scientific literature for eukaryotes.

THE CONCEPTUAL REVOLUTION

Gene regulation instead plays the major role in phenotypic evolution. Both with respect to the ontogenetic differentiation of body parts within an individual and the phylogenetic differentiation of body plans across diverse forms of life, changes in genetic regulation (how structural genes are switched on or off, or rheostated) assume paramount importance during the evolutionary process.

In 1969, geneticists Roy Britten and Eric Davidson proposed a remarkable molecular model for how such gene regulation might transpire in eukaryotes. In their conceptual construct, the authors envisioned batteries of producer loci (protein-coding or other housekeeping genes) being coordinately regulated by activator RNA molecules synthesized by integrator genes whose effect was to induce or repress joint transcription from multiple producer genes in response to environmental stimuli (such as hormones) acting on particular sensor genes. The model remains worthy of study today, not because it has proved to be correct in all its details (it has not), but because it offers a prime example of *avant-garde* thinking in evolutionary genomics. Today, the study of gene

J.C. Avise: Conceptual Breakthroughs in Evolutionary Genetics.
DOI: http://dx.doi.org/10.1016/B978-0-12-420166-8.00041-5

regulation remains one of the hottest topics in evolutionary biology (ENCODE Project Consortium, 2012), with diverse regulatory mechanisms routinely illuminated at transcriptional, translational, and post-translational levels (see Chapter 63).

PS-score: 7

This is an example of a paradigm shift that was gradual in the making. During the 1950s and 1960s, evidence slowly accumulated from a wide variety of sources suggesting the many ways in which gene expression might be regulated, both during ontogeny and across the course of evolutionary time. However, Britten and Davidson's landmark paper in 1969 helped to crystallize many of these emerging ideas. Their classic paper highlighted the potential significance of gene regulation, and also formulated an innovative hypothesis for how such processes might transpire in a mechanistic sense. Their molecular genetic model, which incorporated recent discoveries on the ubiquity of repetitive DNA in eukaryotic genomes (see Chapter 39), was conceptually far ahead of its time. This paradigm shift also gets a high score because nearly all evolutionary geneticists now fully accept the idea that changes in the regulatory apparatus of eukaryotic genomes are central to much of adaptive evolution.

REFERENCES AND FURTHER READING

Jacob F, Monod J. 1961. Genetic regulatory mechanisms in the synthesis of proteins. *J. Mol. Biol.* 3:318–356.

Britten RJ, Davidson EH. 1969. Gene regulation for higher cells: a theory. *Science* 165:349–357.

Raff RA, Kaufman TC. 1983. *Embryos, Genes, and Evolution: The Developmental Genetic Basis of Evolutionary Change*. Macmillan, New York, NY.

Carroll SB, Grenier JK, Weatherbee SD. 2001. *From DNA to Diversity: Molecular Genetics and the Evolution of Animal Design*. Blackwell, London, UK.

ENCODE Project Consortium. 2012. An integrated encyclopedia of DNA elements in the human genome. *Nature* 489:57–74.

Chapter 42

1970
The Flow of
Information

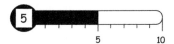

THE STANDARD PARADIGM

In biochemical genetics, the molecular direction of information flow is invariably from DNA → RNA → protein. In other words, DNA is first transcribed into RNA, which then may be translated into polypeptides that make up proteins. This view was so ensconced in the field that it had become known as the "central dogma" (Crick, 1970) of molecular biology.

THE CONCEPTUAL REVOLUTION

Researchers showed that biochemical information could also flow from RNA → DNA. The key discovery came when Howard Temin and David Baltimore, working independently and on different viral systems, identified an enzyme (reverse transcriptase) that catalyzes the conversion of RNA into DNA, thus enabling the passage of genetic information in a direction contrary to the central dogma.

PS-score: 5

Temin and Baltimore shared the 1975 Nobel Prize for their discovery, testifying to the scientific impact of their work. Today, reverse transcriptase is widely used in the biotechnology industry and in molecular genetics laboratories around the world. Reverse transcription is also recognized to play important biological roles in cells. For example, many transposable elements (see Chapter 24) replicatively transpose from one chromosomal site to another via an RNA intermediate that is reverse transcribed; so too do many "processed" pseudogenes (see Chapter 68). Indeed, processed pseudogenes are a major subclass of pseudogenes typically recognizable precisely because

J.C. Avise: Conceptual Breakthroughs in Evolutionary Genetics.
DOI: http://dx.doi.org/10.1016/B978-0-12-420166-8.00042-7

they derive via reverse transcription from a mature messenger RNA from which the intervening sequences or introns (see Chapter 49) already have been spliced out by the cell.

To my knowledge, no one has firmly documented the natural flow of biochemical genetic information from protein→RNA. Thus, at least this portion of the central dogma appears to remain intact. As an interesting footnote to this story, Francis Crick later expressed regret that he had used the word "dogma" in this scientific context, because the word technically means "belief that cannot be doubted" (which by definition would place any dogma outside the purview of science).

REFERENCES AND FURTHER READING

Baltimore D. 1970. RNA-dependent DNA polymerase in virions of RNA tumour viruses. *Nature* 226:1209–1211.

Crick FHC. 1970. Central dogma of molecular biology. *Nature* 227:561–563.

Temin HM, Mizutani S. 1970. RNA-dependent DNA polymerase in virions of Rous sarcoma virus. *Nature* 226:1211–1213.

Weiner AM, Deininger PL, Efstratiadis A. 1986. Nonviral retroposons: genes, pseudogenes, and transposable elements generated by the reverse flow of genetic information. *Annu. Rev. Biochem.* 55:631–661.

Chapter 43

1970
Post-copulatory
Sexual Selection

THE STANDARD PARADIGM

Sexual selection involves differential success in mate acquisition. In his second most influential book, Darwin (1871) introduced the concept of sexual selection via female choice and male—male competition for mates (see Chapter 4). Darwin envisioned both of these processes as operating at the pre-copulatory phase of the reproductive process. In an uncharacteristic oversight, Darwin apparently never contemplated the possibility of post-copulatory sexual selection within the female's reproductive tract.

THE CONCEPTUAL REVOLUTION

A full century later, Darwin's mental lapse was finally rectified when Geoffrey Parker (1970) first compiled evidence for a form of sexual selection that he christened "sperm competition", which entails post-ejaculatory gametic competition for fertilizations (as opposed to direct competition of males for mates). Several years later, Randy Thornhill (1983) coined the phrase "cryptic female choice" to refer to a related post-copulatory process, in which the female reproductive tract plays an active role in deciding the fertilization fate of the sperm (often from multiple males) that were inseminated into her.

PS-score: 5

In the past four decades, the illumination of sexual selection's two post-copulatory mechanisms (sperm competition and cryptic female choice) has produced a plethora of reproductive analyses related to the extraordinary diversity of male and female genitalia, the structural and functional designs

J.C. Avise: Conceptual Breakthroughs in Evolutionary Genetics.
DOI: http://dx.doi.org/10.1016/B978-0-12-420166-8.00043-9

of male and female gametes, the composition of seminal and other reproductive fluids, and copulation behaviors in many organismal groups. For each of these topics, extensive evidence has accumulated — from a wide variety of taxa — for the evolutionary importance and generality of post-copulatory sexual selection. For example, comparisons of testes size and males' rates of sperm production across various taxa have been shown to correlate positively with the presumed intensities of sperm competition as deduced from the genetic mating systems (see Chapter 58) and degrees of polygamy (multiple mating).

REFERENCES AND FURTHER READING

Parker GA. 1970. Sperm competition and its evolutionary consequences in insects. *Biol. Rev.* 45:525–567.

Thornhill R. 1983. Cryptic female choice and its implications in the scorpionfly *Harpobittacus nigriceps*. *Amer. Natur.* 122:765–788.

Smith RL (Ed.). 1984. *Sperm Competition and the Evolution of Animal Mating Systems.* Academic Press, London, UK.

Birkhead TR, Møller AP (Eds). 1998. *Sperm Competition and Sexual Selection.* Academic Press, San Diego, CA.

Birkhead T. 2000. *Promiscuity: An Evolutionary History of Sperm Competition.* Harvard University Press, Cambridge, MA.

Birkhead TR. 2010. How stupid not to have thought of that: post-copulatory sexual selection. *J. Zool.* 281:78–93.

Chapter 44

1972
Jerky Evolution

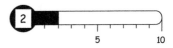

THE STANDARD PARADIGM

Evolution proceeds at a steady, gradual pace. In other words, "phyletic grad-ualism" was thought to be the evolutionary norm for organismal phenotypes, both within and across organismal lineages.

THE CONCEPTUAL REVOLUTION

According to paleontologists Niles Eldredge and Stephen J. Gould, evolution is better characterized by long periods of stasis punctuated by bursts of phe-notypic change typically associated with speciation. This much-discussed model was termed punctuated equilibrium or "rectangular" evolution (the lat-ter coming from the fact that nodes in phylogenetic trees would look squared-off rather than V-shaped if most evolutionary change took place dur-ing speciation events). The model grew from the frequent observation in the fossil record of apparent phenotypic discontinuities from sequential geologi-cal strata. Eldredge and Gould interpreted these discontinuities as being literal footprints of rapid bursts of phenotypic change associated with speciation.

PS-score: 2

Notwithstanding its considerable conceptual impact in the field of paleontol-ogy, I give this purported paradigm shift a relatively low ranking for several reasons. First, in my opinion the authors overstated their case that the tradi-tional evolutionary view entailed strict phyletic gradualism. For example, the paleontologist George Gaylord Simpson — one of the main architects of the modern evolutionary synthesis (see Chapter 21) — had championed the notion that evolution has multiple tempos and modes. Second, the punctuated equilibrium scenario itself morphed through time (beginning as a simple

J.C. Avise: Conceptual Breakthroughs in Evolutionary Genetics.
DOI: http://dx.doi.org/10.1016/B978-0-12-420166-8.00044-0

extension of well-established models of allopatric speciation but later shifting to an emphasis on long-term stasis within an evolutionary lineage). Thus it has proved to be a moving target for its critics. Third, the proposal gave the misleading impression that microevolution (population genetics) and macroevolution were somehow uncoupled, which cannot ultimately be true. Finally, empirical evidence in favor of rectangular evolution (and its insistence that nearly all evolutionary change is compacted into speciation events) remains contentious, at the levels of both genotypes and organismal phenotypes. Thus, overall, this scientific revolution appears to have been rather heavy on hyperbole but somewhat light on substance.

REFERENCES AND FURTHER READING

Simpson GG. 1944. *Tempo and Mode in Evolution.* Columbia University Press, New York, NY.
Eldredge N, Gould SJ. 1972. Punctuated equilibria: an alternative to phyletic gradualism. In: Schopf TJM (Ed.). *Models in Paleobiology.* Freeman, San Francisco, CA, pp. 82–115.
Stanley SM. 1975. A theory of evolution above the species level. *Proc. Natl Acad. Sci. USA* 72:646–650.
Stebbins GL, Ayala FJ. 1981. Is a new evolutionary synthesis necessary? *Science* 213:967–971.
Gould SJ. 2002. *The Structure of Evolutionary Theory.* Harvard University Press, Cambridge, MA.

Chapter 45

1972
Recombinant DNA

THE STANDARD PARADIGM

Natural processes alone engineer pre-existing genetic material. Even as the complicated molecular machinery of life was gradually being illuminated during the middle third of the 20th century (see Chapter 23), most scientists continued to suppose that myriad molecular genetic operations (such as nucleic acid cutting, splicing, and genomic cloning) were under the sole jurisdiction of Mother Nature.

THE CONCEPTUAL REVOLUTION

A revolutionary shift in attitude occurred in the early 1970s, when scientists for the first time began to harness natural enzymatic and other biochemical processes to consciously engineer extant DNA and thereby generate "transgenic" organisms in the laboratory. In other words, molecular geneticists began to employ direct molecular genetic means for the purposeful end of altering pre-existing life-forms. An especially noteworthy development occurred in 1972 when the biochemist Paul Berg and his colleagues reported that they had spliced the DNA of one virus (SV40) into the DNA of another virus (lambda phage) that infects the bacterium *Escherichia coli*. This was the first demonstration of the human-driven *in vitro* splicing of the DNAs from two different organisms. This landmark achievement in the artificial production and manipulation of "recombinant DNA" won Berg a Nobel Prize in Chemistry in 1980.

PS-score: 7

The experiments by Berg and his colleagues in effect marked the birth of genetic engineering and the production of transgenic GMOs (genetically modified organisms). Genetic engineering via recombinant DNA technologies

J.C. Avise: Conceptual Breakthroughs in Evolutionary Genetics.
DOI: http://dx.doi.org/10.1016/B978-0-12-420166-8.00045-2

is now a multibillion-dollar global industry affecting countless endeavors in medicine, agribusiness, animal husbandry, and many other commercial enterprises. It has also conceptually impacted the field of evolutionary genetics by raising the specter of human-mediated evolution of genetically novel lifeforms. Genetic engineering is certainly not without its critics, because recombinant DNA technologies have many pitfalls as well as promises. Interestingly, Berg himself was quite concerned about the potential dangers of transgenic organisms, as evidenced by the fact that in 1973 he convened an international conference to discuss biohazard issues and safety procedures for this new molecular genetic licence. Today, such regulation often tends to be much more formalized. For example, in the United States three federal agencies – The Department of Agriculture (USDA), the Food and Drug Administration (FDA), and the Environmental Protection Agency (EPA) – help to regulate a GMO agri-industry that produces a huge reservoir of genetically engineered crops and transgenic food products.

REFERENCES AND FURTHER READING

Jackson DA, Symons RH, Berg P. 1972. Biochemical method for inserting new genetic information into DNA of simian virus 40: circular SV40 DNA molecules containing lambda phage genes and the galactose operon of *Escherichia coli. Proc. Natl Acad. Sci. USA* 69:2904–2909.

Lyon J, Gorner P. 1995. *Altered Fates: Gene Therapy and the Retooling of Human Life*. Norton & Co., New York, NY.

Stock G, Campbell J. (Eds). 2000. *Engineering the Human Germline*. Oxford University Press, New York, NY.

Avise JC. 2004. *The Hope, Hype, and Reality of Genetic Engineering*. Oxford University Press, Oxford, UK.

Chapter 46

1974 Parent–Offspring Conflict

THE STANDARD PARADIGM

Parents and their progeny share a mutuality of evolutionary interests that fosters behavioral and genetic cooperation between them. This assumption seemed so logical that it was seldom challenged in the evolutionary literature.

THE CONCEPTUAL REVOLUTION

In 1974, evolutionary biologist Robert Trivers published a paper showing that this conventional expectation is an oversimplification at best; strategic genetic conflicts between parent and child can be rampant even in such sacrosanct locations and times as within the mammalian womb during a pregnancy. For example, many physiological conflicts and accommodations exist between a pregnant mother and her sexually produced fetus. Generally, each self-serving offspring is under selection to seek more maternal resources than its mother might wish to relinquish, given the negative effects that such donations can have on a dam's expectation of producing future offspring and thereby on her lifetime genetic fitness. Ultimately, such conflicts are resolved as evolutionary compromises between the oft-competing genetic interests of mother and child. Proximately, maternal–fetal conflicts during a pregnancy can register as a wide array of health problems ranging from the subtle (such as morning sickness) to the egregious (e.g., fetal abortion). Conflicts of genetic interest between parent and child also extend into the post-partum realm and to father–offspring relations. The bottom line is that parent–offspring interactions are not always the product of an adaptive, harmonious relationship between parent and offspring, but instead are shaped by inherent evolutionary–genetic conflicts between fathers, mothers, and their progeny.

J.C. Avise: Conceptual Breakthroughs in Evolutionary Genetics.
DOI: http://dx.doi.org/10.1016/B978-0-12-420166-8.00046-4

PS-score: 4

The classic article by Trivers provides an important example of a broader paradigm shift during the 1970s toward analyzing social biological phenomena from a Darwinian evolutionary perspective (see Chapter 36). As such, it also gave added impetus to the emerging fields of sociobiology (see Chapter 32) and coevolution (see Chapter 33).

REFERENCES AND FURTHER READING

Trivers RL. 1974. Parent–offspring conflict. *Amer. Zool.* 14:249–264.

Trivers RL. 1985. *Social Evolution*. Benjamin-Cummings, Menlo Park, CA.

Clutton-Brock TH. 1991. *The Evolution of Parental Care*. Princeton University Press, Princeton, NJ.

Haig D. 1993. Genetic conflicts in human pregnancy. *Q. Rev. Biol.* 68:495–532.

Avise JC. 2013. *Evolutionary Perspectives on Pregnancy*. Columbia University Press, New York, NY.

Chapter 47

1975
Human Genomic
Uniqueness

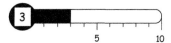

THE STANDARD PARADIGM

Humans are extraordinarily distinct from all other creatures. Throughout antiquity and continuing into the modern era, a nearly universal sentiment had been that humans occupy a totally unique pinnacle in the biological world. Although Darwin certainly had challenged this notion with his insights about the evolutionary process and humanity's place in nature, the paradigm remained that human beings are genetically nonpareil.

THE CONCEPTUAL REVOLUTION

In 1975, any lingering sentiment that the genome of *Homo sapiens* might be qualitatively matchless were put to rest when Mary-Claire King and Allan Wilson famously summarized available genetic findings showing that protein and DNA sequences in modern humans are more than 99% identical to their counterpart macromolecules in chimpanzees. King and Wilson's classic paper also showcased the novel idea that evolution can proceed at different tempos and modes at the levels of genotype versus phenotype, which in turn fed into the notion that changes in genetic regulation might be at the heart of phenotypic evolution (see Chapter 41). Despite their many phenotypic differences, humans and chimpanzees are remarkably alike at the elemental level of their macromolecular sequences. Of course, with the benefit of hindsight, this makes considerable sense in the light of evolution, given the relative recency of the phylogenetic split between the proto-chimp and the proto-human lineages only about six million years ago. Not enough time has elapsed since shared ancestry for greater genetic differences to have accumulated between humans and the other living great apes.

J.C. Avise: Conceptual Breakthroughs in Evolutionary Genetics.
DOI: http://dx.doi.org/10.1016/B978-0-12-420166-8.00047-6

PS-score: 3

This paradigm shift would be scored higher except for the fact that its basic sentiment had been anticipated by many biologists (at least since Darwin), as well as by many philosophers and social scientists, who sometimes have viewed humans as closely allied to primates and other mammals. And, of course, there can be little doubt that *Homo sapiens* is indeed very special with regard to countless organic features, including its neurobiology, mental consciousness, capacity for language, and many other genetic endowments that ultimately underlie the uniquely human dispositions for culture and for social evolution.

REFERENCES AND FURTHER READING

King M-C, Wilson AC. 1975. Evolution at two levels in humans and chimpanzees. *Science* 188:107–116.

Cherry LM, Case SM, Wilson AC. 1978. Frog perspective on the morphological divergence between humans and chimpanzees. *Science* 200:209–211.

Carroll SB. 2003. Genetics and the making of *Homo sapiens. Nature* 422:849–857.

Patterson N, Richter DJ, Gnerre S, Lander ES, Reich D. 2006. Genetic evidence for complex speciation of humans and chimpanzees. *Nature* 441:315–321.

Cela-Conde C, Ayala FJ. 2007. *Human Evolution: Trails from the Past.* Oxford University Press, Oxford, UK.

Avise JC, Ayala FJ. 2010. *In the Light of Evolution, IV: The Human Condition.* National Academies Press, Washington, DC.

Prado-Martinez J, Sudmant PH, Kidd JM, Li H, Kelley JL, Lorente-Galdos B., et al., 2013. Great ape genetic diversity and population history. *Nature* 499:471–475.

Chapter 48

1976
Selfish Genes

THE STANDARD PARADIGM

Most genes within any genome functionally collaborate in harmonious fashion on behalf of organismal fitness. With respect to producing functional and healthy organisms, harmonious interaction (adaptive epistasis; see Chapter 10) was expected among most genetic loci in any species. In other words, most genes were expected to collaborate in positive ways on behalf of improving organismal fitness.

THE CONCEPTUAL REVOLUTION

Many genes behave instead as if acting in their own self-serving interests. If true, the level of selection would often be at the level of the gene as much or more so than at the level of the organism (see Chapter 36) or the population.

PS-score: 9

Sometimes a scientific revolution is spearheaded not by a singular discovery but rather by a powerful argument eloquently presented. Apart from Darwin's treatise on natural selection, I can think of no better example than the concept of the selfish gene as elaborated in a popular book authored by the renowned evolutionary biologist Richard Dawkins. In that highly influential work, Dawkins developed the then-unorthodox thesis that genetic loci in sexual species seem to have invented (evolved) many ingenious ways to spread and persist in populations without necessarily contributing in a positive way to organismal fitness. Today, examples of such self-serving tactics by various genetic loci are known to include: interference, in which a selfish allele disrupts or sabotages the transmission of alternate alleles at the same genetic locus; mechanisms by which selfish DNA moves preferentially toward the germline (where it will more likely be transmitted to the next

J.C. Avise: Conceptual Breakthroughs in Evolutionary Genetics.
DOI: http://dx.doi.org/10.1016/B978-0-12-420166-8.00048-8

generation); and overreplication, in which a segment of selfish DNA biases its intergenerational transmission by getting itself replicated more often than other loci in the same host. Transposable elements (see Chapter 24) provide a quintessential example of the latter category of selfish genes. These ubiquitous features of sexual genomes replicatively proliferate across multiple chromosomal locations and thereby greatly improve their chances for successful passage to the next organismal generation.

The powerful image of the selfish gene ranks high on my list of paradigm busters because it fundamentally and almost universally altered scientific opinion about the hierarchical level at which natural selection can act. Natural selection at the level of the gene is not incompatible with natural selection at the level of the individual (see Chapter 36) or kinship group (see Chapter 32), but rather it offers an embellished view of the entire selective process. The revolutionary concept of the selfish gene is now widely acknowledged to be a key paradigm in evolutionary genetics.

REFERENCES AND FURTHER READING

Dawkins R. 1976. *The Selfish Gene*. Oxford University Press, Oxford, UK.
Burt A, Trivers R. 2006. *Genes in Conflict: The Biology of Selfish Genetic Elements*. Harvard University Press, Cambridge, MA.

Chapter 49

1977
Split Genes

THE STANDARD PARADIGM

Each gene is an uninterrupted sequence of nucleotides specifying a functional polypeptide or RNA molecule. This sentiment was thought to be especially true for "good-housekeeping" loci encoding the proteins that are involved in central metabolic pathways and other cell-housekeeping chores.

THE CONCEPTUAL REVOLUTION

Philip Sharp and Richard Roberts, working independently, discovered that many genes are actually split or interrupted by what soon became known as intervening sequence regions or "introns" (Gilbert, 1978) (in contradistinction to expressed sequences or "exons" that encode the functional products of a gene). Further research showed that intronic sequences in transcribed RNA are subsequently removed during a process known as RNA splicing. During this cellular operation, mediated by a complex biochemical apparatus known as a spliceosome, introns are cleaved out of a pre-messenger RNA, and exons are stitched back together in proper order (or sometimes in a different order) before undergoing translation from the final mature messenger (m) RNA product. After the spliceosomes have finished their work, the mature mRNA molecules are exported to the cell cytoplasm where they serve as templates for protein translation on ribosomes.

PS-score: 7

The importance of introns is amplified by the fact that they have now been found in a wide variety of genes from diverse organisms ranging from bacteria and other microbes to humans. (For example, the human genome carries an average of 8.4 introns per gene.) With regard to their evolutionary significance, many roles for introns have been proposed, and in some cases well

J.C. Avise: Conceptual Breakthroughs in Evolutionary Genetics.
DOI: http://dx.doi.org/10.1016/B978-0-12-420166-8.00049-X

documented. For example, alternative splicing permits great developmental flexibility and exon shuffling permits great evolutionary flexibility in how proteins are constructed from their constituent genic parts. A very different scenario (albeit not mutually exclusive to the first) is that some introns are mobile elements that epitomize selfish DNA (see Chapter 48) and serve no particular cellular function apart from their own self-perpetuation. In 1993, Sharp and Roberts were awarded a Nobel Prize for their revolutionary discovery of split genes.

REFERENCES AND FURTHER READING

Berget SM, Moore C, Sharp PA (principal investigator). 1977. Spliced segments at the 5′ terminus of adenovirus 2 late mRNA. *Proc. Natl Acad. Sci. USA* 74:3171–3175.

Chow LT, Gelinas RE, Broker TR, Roberts RJ (principal investigator). 1977. An amazing sequence arrangement at the 5′ ends of adenovirus 2 messenger RNA. *Cell* 12:1–8.

Gilbert W. 1978. Why genes in pieces? *Nature* 271:501.

Rodríguez-Trelles F, Tarrío R, Ayala FJ. 2006. Origins and evolution of spliceosomal introns. *Annu. Rev. Genet.* 40:47–76.

Chapter 50

1977
Domains of Life

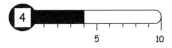

THE STANDARD PARADIGM

Life comes in two basic flavors: prokaryotic and eukaryotic. For decades, textbooks in evolution and genetics had preached that life on Earth was divisible into two great kingdoms (each presumably monophyletic): organisms whose cells contain a membrane-enclosed nucleus (the Eukaryotes, including plants, animals, and fungi), versus organisms whose cells lack membrane-bound nuclei (the Prokaryotes or bacteria).

THE CONCEPTUAL REVOLUTION

A fundamental biological fact is that all cellular organisms have ribosomes — multisubunit molecular organelles that are made up of ribosomal (r) RNA subunits and proteins that collectively orchestrate protein synthesis by translating messenger RNA (mRNA) molecules into polypeptides. Because of their ubiquity across cellular life forms, ribosomal RNA sequences offer a potentially universal phylogenetic marker. Based on this startling realization, in 1977 Carl Woese and George Fox compared rRNA sequences from many microbes and other creatures, and the results led them to challenge conventional taxonomic wisdom by suggesting that life is split into three aboriginal phylogenetic domains: the Eubacteria (later called Bacteria), comprising all typical bacteria; the Archaebacteria (later called Archaea), containing methanogenic bacteria and other "extremophiles" (many of which live in seemingly ultra-hostile environments such as hot springs and hypersaline brine pools); and the Urkaryotes (or Eukarya), as now represented by traditional eukaryotes. Thus a deep phylogenetic crevice — previously unrecognized — apparently sunders the microbial world into two distinct historical subsections. In other words, the former prokaryotic arena seemed to be comprised of two primordial phylogenetic lineages, such that altogether at least three primary phylogenetic domains of life exist. This rearrangement of the tree of

J.C. Avise: Conceptual Breakthroughs in Evolutionary Genetics.
DOI: http://dx.doi.org/10.1016/B978-0-12-420166-8.00050-6

life raised many questions that even today remain somewhat problematic: Are eubacteria, archaea, and eukaryotes each fully monophyletic? What are the branch lengths leading to extant forms? What is the correct rooting of the universal phylogenetic tree? Are there other ancient but cryptic microbial lineages? To which branch of the microbial world are the eukaryotes related most closely?

PS-score: 4

Although the revelation by Woese and Fox provoked a negative response by many microbiologists at that time, its central tenet is now accepted almost universally. Nonetheless, this phylogenetic discovery gets only a median score on my *PS-index*. In its favor is the fact that it has rewritten the text-books and generally withstood the test of time. In its disfavor is the fact that phylogeneticists had long used protein and DNA sequences to make count-less discoveries about organismal relationships at multiple levels in the Linnean hierarchy (see Chapter 31). Thus, what really sets the Woese and Fox study apart from myriad other molecular phylogenetic analyses is not so much its conceptual innovation but rather the grandiose phylogenetic scale of its findings.

In 1992, the Royal Netherlands Academy of Arts and Sciences awarded Carl Woese the Leeuwenhoek Medal, microbiology's premiere honor.

REFERENCES AND FURTHER READING

Woese CR, Fox GE. 1977. Phylogenetic structure of the prokaryotic domain: the primary king-doms. *Proc. Natl Acad. Sci. USA* 74:5088–5090.

Rinke C, Schwientek P, Sczyrba A, Ivanova NN, Anderson IJ, Cheng J-F, et al., 2013. Insights into the phylogeny and coding potential of microbial dark matter. *Nature* 499:431–437.

Chapter 51

1979
Phylogeography

THE STANDARD PARADIGM

Phylogeny is a supraspecific concept and has no meaning at the intraspecific level. Evolutionary genetics traditionally was subdivided into two major sub-disciplines: phylogenetics above the species level, and population genetics below the species level. Speciation was widely viewed as a demarcation line across which phylogeneticists dare not tread because intraspecific evolution was within the realm of lineage reticulation (anastomosis) mediated by sexual reproduction (see Chapter 20). Thus, evolutionary trees and phylogenetic reasoning purportedly did not apply within species.

THE CONCEPTUAL REVOLUTION

In the late 1970s, the introduction of mitochondrial (mt) DNA (see Chapter 13) to population biology demonstrated that the non-reticulate matrilineal component of organismal phylogeny could be empirically recovered within and among conspecific populations. To capture this new viewpoint on intraspecific genealogy, the word phylogeography was coined in 1987. Phylogeographic perspectives have a series of corollaries, each of which could warrant designation as a substantial evolutionary paradigm shift in its own right. Some of these new perspectives are as follows: (1) many quasi-distinct gene trees exist in any lineage of sexual reproducers, and these gene trees are conceptually recognizable components of the organismal phylogeny; (2) historical population demography and intraspecific phylogeny are inextricably intertwined, because means and variances in offspring numbers dictate the structures of gene genealogies; (3) individuals (as well as populations) can properly be treated as "operational taxonomic units" (OTUs) in genealogical appraisals at the intraspecific level; and (4) microevolution, like macroevolution, is inherently historical. Phylogeography also helped to spawn coalescent theory, which traces lineages back to shared ancestors, and

J.C. Avise: Conceptual Breakthroughs in Evolutionary Genetics.
DOI: http://dx.doi.org/10.1016/B978-0-12-420166-8.00051-8
107

which has become a major branch of mathematical and statistical theory at the boundary between population genetics and phylogenetic biology.

PS-score: 8

It is hard for me to remain objective about this paradigm shift because I was one of its primary architects. Nevertheless, I think phylogeography merits a high *PS-score* because it has spawned a host of corollary paradigm shifts, it has become widely accepted by the scientific community, and it is difficult to imagine returning to the earlier era when population genetics (microevolution) and phylogenetics (macroevolution) were widely considered to be empirically and conceptually disconnected.

REFERENCES AND FURTHER READING

Avise JC, Giblin-Davidson C, Laerm J, Patton JC, Lansman RA. 1979. Mitochondrial DNA clones and matriarchal phylogeny within and among geographic populations of the pocket gopher, *Geomys pinetis*. *Proc. Natl Acad. Sci. USA* 76:6694–6698.

Avise JC, Lansman RA, Shade RO. 1979. The use of restriction endonucleases to measure mitochondrial DNA sequence relatedness in natural populations. I. Population structure and evolution in the genus *Peromyscus*. *Genetics* 92:279–295.

Avise JC, Arnold J, Martin Ball R, Bermingham E, Lamb T, Neigel JE, et al., 1987. Intraspecific phylogeography: the mitochondrial DNA bridge between population genetics and systematics. *Annu. Rev. Ecol. Syst.* 18:489–522.

Avise JC. 1989. Gene trees and organismal histories: a phylogenetic approach to population biology. *Evolution* 43:1192–1208.

Hudson RR. 1991. Gene genealogies and the coalescent process. *Oxford Surv. Evol. Biol.* 7:1–44.

Avise JC. 2000. *Phylogeography: The History and Formation of Species*. Harvard University Press, Cambridge, MA.

Wakeley J. 2010. *Coalescent Theory: An Introduction*. Roberts & Co, Chicago, IL.

Nielsen R, Slatkin M. 2013. *An Introduction to Population Genetics: Theory and Applications*. Sinauer, Sunderland, MA.

Chapter 52

1979
Exaptations

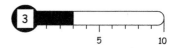

THE STANDARD PARADIGM

Nearly all phenotypic features can be understood as adaptations to particular evolutionary challenges. Gould and Lewontin (1979) derisively referred to this dogma as the Panglossian paradigm of the adaptationist programme in evolutionary biology.

THE CONCEPTUAL REVOLUTION

According to Stephen Gould and Richard Lewontin, many organismal traits are functionally suboptimal because they may have evolved originally to serve entirely different functions (or perhaps no particular function at all). An example might be the feathers that now support avian flight but were selectively favored at the outset for their role in thermoregulation. Another example might involve the lungs of primitive fish that eventually evolved into a gas bladder that now regulates buoyancy in modern fish. Yet another example might be a mobile element (see Chapter 24) that has proliferated for selfish reasons (see Chapter 48) but later became coopted by the cell for some useful role in genetic regulation for the host organism.

In 1982, Gould and Vrba coined the term "exaptation" to refer to situations in which: (1) a trait previously shaped by natural selection for a particular function is coopted for another use; or (2) a trait whose origin cannot be ascribed to the direct action of natural selection later becomes coopted for its current utility. The authors viewed a focus on exaptations to be a useful and needed conceptual counterbalance to standard adaptive explanations for many organismal features. In other words, phenotypic evolution involves a lot of what Francois Jacob (1977) had called evolutionary tinkering.

J.C. Avise: Conceptual Breakthroughs in Evolutionary Genetics.
DOI: http://dx.doi.org/10.1016/B978-0-12-420166-8.00052-X

PS-score: 3

This new paradigm gets high marks for its veracity and for the eloquence with which it was presented by Gould and Lewontin, but much lower marks for its level of originality. Long before the term exaptation was introduced, evolutionary biologists routinely used the term preadaptation to refer to traits that had been coopted during evolution for some biological function other than that provided by the original. The word preadaptation became outmoded, however, because it could be misinterpreted to imply the operation of some sort of teleological process that would be contrary to evolutionary principles.

REFERENCES AND FURTHER READING

Schmalhausen II. 1949. *Factors of Evolution. The Theory of Stabilizing Selection.* Blakiston, Philadelphia, PA.

Jacob F. 1977. Evolution and tinkering. *Science* 196:1161–1166.

Gould SJ, Lewontin RC. 1979. The spandrels of San Marco and the panglossian paradigm: a critique of the adaptationist programme. *Proc. R. Soc. Lond. B.* 205:581–598.

Gould SJ, Vrba E. 1982. Exaptation – a missing term in the science of form. *Paleobiology* 8:4–15.

Buss DM, Haselton MG, Shackelford TK, Bleske AL, Wakefield JC. 1998. Adaptations, exaptations, and spandrels. *Amer. Psychol.* 53:533–548.

Chapter 53

1979
Genetic Code

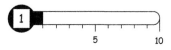

THE STANDARD PARADIGM

The genetic code is universal. The genetic code is the set of biochemical rules governing how genetic information encoded in nucleic acids is converted into polypeptides (proteins). Basically, in a translational process that takes place on a cell's ribosomes, successive triplets of nucleotides in a nucleic acid specify precisely which amino acids are to be added to a growing polypeptide chain (see Chapter 42). For example, the triplet AAA in an RNA sequence normally specifies the incorporation of a lysine, whereas the triplet CGC specifies arginine. With four different types of nucleotides (A, T, C, and G), a total of $(4)^3 = 64$ different nucleotide triplets exist. The genetic code typically is presented as a table showing which amino acid is specified by each such codon. The genetic code was painstakingly deciphered in the early 1960s (an accomplishment for which Har Khorana, Robert Holley, and Marshall Nirenberg later shared a Nobel Prize), and for the next two decades it was assumed to apply universally to all forms of life on Earth.

THE CONCEPTUAL REVOLUTION

In 1979, researchers discovered exceptions that broke the standard coding rules. Specifically, mammalian mitochondrial (mt) DNA was shown to depart from the "universal" genetic code in several molecular details. Subsequently, additional variations in the genetic code were discovered among mitochondrial genomes of various animal taxa, fungi, and plants, and in the genomes of protozoans and bacteria.

PS-score: 1

Although this paradigm shift was both unambiguous and unanticipated, it has had limited impact on the field of evolutionary genetics, apart from

J.C. Avise: Conceptual Breakthroughs in Evolutionary Genetics.
DOI: http://dx.doi.org/10.1016/B978-0-12-420166-8.00053-1

stimulating interesting conversations on how changes in the genetic code might have transpired during the evolutionary process (see, for example, Osawa et al., 1992). Because alterations in the genetic code are individually rare events in evolution, in some cases they have also proved to be useful phylogenetic markers of particular organismal clades.

REFERENCES AND FURTHER READING

Crick FHC. 1968. The origin of the genetic code. *J. Mol. Biol.* 38:367–379.

Barrell BG, Bankier AT, Drouin J. 1979. A different genetic code in human mitochondria. *Nature* 282:189–194.

Jukes TH. 1983. Mitochondrial codes and evolution. *Nature* 301:19–20.

Fox TD. 1985. Diverged genetic codes in protozoans and a bacterium. *Nature* 314:345–376.

Osawa S, Jukes TH, Wanatabe K, Muti A. 1992. Recent evidence for evolution of the genetic code. *Microbiol. Rev.* 56:229–264.

Wolstenholme DR. 1992. Animal mitochondrial DNA: structure and evolution. *Int. Rev. Cytol.* 141:173–216.

Post-1980: Elaborating and Revisiting the Foundations

The conceptual breakthroughs described in Part IV include a potpourri of recent topics, beginning with the discovery that many microbial populations are predominantly clonal rather than freely recombining, that some RNAs can serve as organic catalysts, and that genetic mating systems in nature routinely differ from the social mating systems as observed directly in the field. Other notable achievements during this modern era were the invention and deployment of evolutionary game theory; elaboration of the concepts and principles of conservation genetics; the widespread use of phylogenetic character mapping; the introductions of Darwinian medicine, evolutionary psychology, and DNA barcoding; the discovery of frequent lateral genetic transfer across evolutionary lineages; and the realization that purported pseudogenes often play functional (e.g., regulatory) roles in cells. Finally, the conceptual breakthroughs of this era came full-circle (in Chapter 69) with the resurrection of Darwin's original sentiment (1859) that natural selection often drives the origin of biological species. Interestingly, only one seminal breakthrough is recognized post-2004, perhaps because insufficient time has yet elapsed to offer a proper temporal perspective on what eventually will be deemed the most impactful achievements among the many scientific contenders during the past decade. In any event, all of the conceptual breakthroughs post-1980, when added to the foundations laid in the first century following Darwin, have eventuated in today's field of modern evolutionary genetics. One can only imagine what the discipline will look like in another 50, 100, or 150 years!

Chapter 54

1980
Microbial Clonality

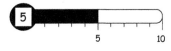

THE STANDARD PARADIGM

In microorganisms, each taxonomic "species" is genetically homogeneous. Mechanisms of genetic mixis, including transformation (the uptake of foreign DNA by a cell), transduction (viral mediated movement of DNA between cells), and conjugation (intercellular mating), were well-documented phenomena in bacteria, so many microbiologists assumed that each bacterial taxon was a genetically well-mixed population. Similarly, in diploid parasitic protozoans a capacity for sexual reproduction had been demonstrated in the laboratory, so many protozoologists likewise assumed that each taxon of these unisexual microbes was genetically well mixed. In other words, most unicellular microbes were thought to be, in effect, primarily sexual rather than predominantly clonal.

THE CONCEPTUAL REVOLUTION

This sentiment began to change in the early 1980s when Robert Selander and his colleagues used genetic markers to demonstrate that populations of the bacterium *Escherichia coli* exist in nature as a series of clonal lineages between which rates of genetic recombination are characteristically low. These discoveries soon were paralleled by similar genetic analyses conducted by Francisco Ayala and colleagues on a wide variety of unicellular micropathogens. As a result of these studies, a new paradigm emerged that many microbes are predominantly clonal in nature. Two genetic footprints of such clonality are strong linkage disequilibrium (non-random allelic associations) across loci and genetic clustering or "near-clading" of microbial lineages in phylogenetic reconstructions of population samples. These phenomena have now been demonstrated in a wide variety of microbial taxa.

J.C. Avise: Conceptual Breakthroughs in Evolutionary Genetics.
DOI: http://dx.doi.org/10.1016/B978-0-12-420166-8.00054-3

PS-score: 5

This paradigm shift is significant because it conceptually transformed the field of microbiology, which previously was one of the last bastions of biology to fully adopt evolutionary genetic reasoning. The paradigm shift was also notable because the population genetic structures of pathogenic and other microbes have many practical implications for medicine, epidemiology, pharmacology, and microbial taxonomy.

REFERENCES AND FURTHER READING

Selander RK, Levin BR. 1980. Genetic diversity and structure in *Escherichia coli* populations. *Science* 210:545–547.

Whittam TS, Ochman H, Selander RK. 1983. Multilocus genetic structure in natural populations of *Escherichia coli*. *Proc. Natl Acad. Sci. USA* 80:1751–1755.

Tibayrenc M, Ward P, Moya A, Ayala FJ. 1986. Natural populations of *Trypanosoma cruzi*, the agent of Chagas disease, have a complex multiclonal structure. *Proc. Natl Acad. Sci. USA* 83:115–119.

Tibayrenc M, Kjellberg F, Ayala FJ. 1990. A clonal theory of parasitic protozoa: the population structures of *Entamoeba*, *Giardia*, *Leishmania*, *Naegleria*, *Plasmodium*, *Trichomonas*, and *Trypanosoma* and their medical and taxonomical consequences. *Proc. Natl Acad. Sci. USA* 87:2414–2418.

Tibayrenc M, Kjellberg F, Arnaud J, Oury B, Brénière SF, Darde ML, et al., 1991. Are eukaryotic microorganisms clonal or sexual? A population genetics vantage. *Proc. Natl Acad. Sci. USA* 88:5129–5133.

Halkett F, Simon J-C, Balloux F. 2005. Tackling the population genetics of clonal and partially clonal organisms. *Trends Ecol. Evol.* 20:194–201.

Chapter 55

1982
Catalyzing New
Thoughts

THE STANDARD PARADIGM

Proteins are the only biological macromolecules capable of catalyzing bio-chemical reactions. Specifically, a subset of proteins known as enzymes were thought to be the world's only natural organic catalysts.

THE CONCEPTUAL REVOLUTION

Researchers showed that RNA molecules could also act as biological cata-lysts by facilitating biochemical reactions under some circumstances. This unexpected discovery was made, independently, in the laboratories of Thomas Cech and Sidney Altman, who in 1989 won the Nobel Prize for their efforts. RNA molecules that can serve catalytic roles quickly became known as ribozymes, in obvious reference and contradistinction to proteinaceous enzymes.

PS-score: 5

Ribozymes subsequently were shown to facilitate a variety of biochemical reactions or functions, such as RNA splicing, viral replication, and the biosyn-thesis of transfer RNA (tRNA). At least one proposed function — regarding origins of life on Earth several billion years ago (see Chapter 27) — carries special relevance for the field of evolutionary genetics. The basic hypothesis (to which many biologists now subscribe) is that RNA may have been both a substrate and an organic catalyst during the very early evolution of pre-biotic self-replicating systems, several billion years ago.

J.C. Avise: Conceptual Breakthroughs in Evolutionary Genetics.
DOI: http://dx.doi.org/10.1016/B978-0-12-420166-8.00055-5

REFERENCES AND FURTHER READING

Kruger K, Grabowski PJ, Zaug AJ, Sands J, Gottschling DE, Cech TR (principal investigator). 1982. Self-splicing RNA – autoexcision and autocyclization of the ribosomal-RNA intervening sequence of *Tetrahymena*. *Cell* 31:147–157.

Guerrier-Takada C, Gardiner K, Marsh T, Pace N, Altman S (principal investigator). 1983. The RNA moiety of ribonuclease-P is the catalytic subunit of the enzyme. *Cell* 35:849–857.

Zaug AJ, Cech TR. 1986. The intervening sequence RNA of *Tetrahymena* is an enzyme. *Science* 231:470–475.

Maynard Smith J, Szathmary E. 1995. *The Major Transitions in Evolution*. Freeman, New York, NY.

Chapter 56

1982
Game Theory

THE STANDARD PARADIGM

Intuitive verbal arguments are sufficient to describe evolutionary outcomes between behaviorally interacting organisms. Sociobiologists (see Chapter 32) focus on the behavioral strategies employed by socially interacting individuals, but before the deployment of game theory there was no cohesive framework for interpreting the long-term (evolutionary) outcomes of alternative behavioral tactics. Game theory can be defined as the study of strategic decision-making. Traditionally used in economics, political science, and psychology, it entails development of mathematical models describing which among two or more competing behavioral tactics yields the greatest payoff (e.g., money) to its practitioners. In human social behavior, applications include bargaining encounters, justice, ultimatum games, commitment choices, decisions about mutual aid versus defection, hawk–dove interactions, ownership issues, and truth versus deception in communications and signaling.

THE CONCEPTUAL REVOLUTION

The application of game theory in an evolutionary context was introduced to biology in the 1970s (although some earlier treatments also exist), and was popularized in a landmark book by John Maynard Smith in 1982 (an achievement for which he later won a Crafoord Prize). This mathematical construct enabled biologists to quantitatively analyse and thereby evaluate competing behavioral tactics with regard to their ultimate payoff, which in the context of evolutionary biology is measured in terms of relative genetic fitness.

PS-score: 7

Game theory continues to find many applications in an animal's "decision-making" in nature. For example, should a family produce mostly sons, mostly

J.C. Avise: Conceptual Breakthroughs in Evolutionary Genetics.
DOI: http://dx.doi.org/10.1016/B978-0-12-420166-8.00056-7

daughters, or some mixed tactic (see Chapter 18)? The usual goal in game theory is to illuminate behavioral tactic(s) that benefit personal fitness, and then to discover whether and under what circumstances particular tactics or combinations of tactics are evolutionarily stable (immune to invasion by alternative tactics) in a population. Each non-invasible outcome is an evolutionary stable strategy or ESS (i.e., the situation toward which a population tends to evolve if the assumptions underlying the model are correct).

REFERENCES AND FURTHER READING

Maynard Smith J. 1982. *Evolution and the Theory of Games*. Cambridge University Press, Cambridge, UK.

Skyrms B. 1996. *Evolution of the Social Contract*. Cambridge University Press, Cambridge, UK.

Maynard Smith J. 1998. *Evolutionary Genetics*, 2nd edition. Oxford University Press, Oxford, UK.

Osborne MJ. 2004. *An Introduction to Game Theory*. Oxford University Press, New York, NY.

Chapter 57

1983 Conservation Genetics

THE STANDARD PARADIGM

Genetic considerations are of relatively minor importance in conservation biology. Despite scientists' newly found access to abundant molecular genetic variation in natural populations (see Chapter 37), the general sentiment remained that conservation biology and the ongoing global extinction crisis were concerns to be addressed primarily by ecologists and natural historians. Genetic input into most conservation deliberations was negligible.

THE CONCEPTUAL REVOLUTION

This situation changed dramatically in 1983 with the publication of *Genetics and Conservation* edited by Christine Schonewald-Cox and her colleagues. This was to be the first of several breakthrough textbooks that soon opened everyone's eyes to the importance of genetic issues (such as inbreeding depression) for endangered and threatened populations. Two general genetic foci eventually emerged: (1) concerns about inbreeding and reduced genetic variation especially in captive (zoo) populations; and (2) the use of molecular markers to assess a wide variety of genetic topics (such as gene flow, population structure, hybridization, and systematics) in natural populations of countless rare species.

PS-score: 7

This is an example of a paradigm shift that can be traced quite directly to the publication of one timely and seminal textbook. The treatment by Schonewald-Cox and coeditors was catholic in its coverage of a wide variety of genetic topics germane to conservation efforts for both captive and wild

J.C. Avise: Conceptual Breakthroughs in Evolutionary Genetics.
DOI: http://dx.doi.org/10.1016/B978-0-12-420166-8.00057-9

populations. In its conceptual breadth, the book was considerably ahead of its time. Today, most biologists appreciate not only that we are in the midst of the sixth mass extinction event (the only one caused by a biological agent — humans) in the planet's long history, but also that genetic issues must be integrated with ecological, demographic, and other considerations in our efforts to manage the many species and populations that are negatively impacted in the unfolding biodiversity crisis.

REFERENCES AND FURTHER READING

Schonewald-Cox CM, Chambers SM, MacBryde B, Thomas L. (Eds) 1983. *Genetics and Conservation*. Benjamin-Cummings, Menlo Park, CA.

Avise JC, Hamrick JL. (Eds) 1996. *Conservation Genetics: Case Histories from Nature*. Chapman & Hall, New York, NY.

Smith TB, Wayne RK. (Eds) 1996. *Molecular Genetic Approaches in Conservation*. Oxford University Press, New York, NY.

Frankham R, Ballou JD, Briscoe DA. 2002. *Introduction to Conservation Genetics*. Cambridge University Press, Cambridge, UK.

Avise JC, Hubbell SP, Ayala FJ. (Eds) 2008. *II: Biodiversity and Extinction*. National Academies Press, Washington, DC.

DeWoody JA, Bickham JW, Michler CH, Nichols KM, Rhodes Jr, OE, Woeste KE. (Eds). 2010. *Molecular Approaches in Natural Resource Conservation and Management*. Cambridge University Press, New York, NY.

Allendorf FW, Luikart G, Aitken SN. 2013. *Conservation and the Genetics of Populations*, 2nd Edition. Wiley-Blackwell, New York, NY.

Chapter 58

1984
DNA Fingerprinting
and Mating Systems

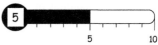

THE STANDARD PARADIGM

Mating systems are what they seem. Mating systems (such as monogamy, polygyny, polyandry, polygynandry, and promiscuity) traditionally were assessed by field observations of who mated with whom (or, in plants, by monitoring pollinator movements). This kind of information was difficult to obtain (especially for secretive species) and at best yielded only the "social mating system" of a population. Ideally, however, researchers would like to know the genetic mating system because this would reveal who actually had mated with whom to transmit genes to the next generation. In the 1960s, the deployment of protein-electrophoretic methods (see Chapter 37) for the first time enabled evolutionary biologists to begin to assess genetic parentage (maternity and paternity) based on molecular polymorphisms in nature, and thereby characterize a population's actual genetic mating system. The general approach involved comparing multi-locus genotypes of candidate parents against those of progeny in focal broods, and then accumulating such information for multiple clutches in each population of interest. In 1984, a related breakthrough was the invention of methods for "DNA fingerprinting" based on hypervariable "minisatellite" regions of eukaryotic genomes.

THE CONCEPTUAL REVOLUTION

By the early 1980s, applications of these and other genetic approaches had begun to reveal that the genetic mating systems of animals and plants frequently do not agree with their apparent social mating systems. For example, many pairs of Eastern Bluebirds formerly thought to be monogamous proved, upon genetic examination, to be rearing foster progeny that must have resulted from surreptitious reproductive behaviors (such as polyandrous

J.C. Avise: Conceptual Breakthroughs in Evolutionary Genetics.
DOI: http://dx.doi.org/10.1016/B978-0-12-420166-8.00058-0
123

matings and/or egg dumping by females). As such genetic data accumulated for many avian species, a paradigm shift took place: many birds formerly assumed to be monogamous are in truth highly polygamous. During the 1980s, 1990s, and 2000s, such genetic appraisals were extended to countless species of animals and plants, with the net result being a conceptual revolution in our understanding of genetic mating systems and alternative reproductive tactics in a wide variety of taxa. Today, genetic appraisals of biological maternity and paternity are conducted quite routinely, typically using highly polymorphic microsatellite markers.

PS-score: 5

This shift to a new paradigm (that genetic mating systems in many species differ from social mating systems because of surreptitious reproductive behaviors) was gradual and is still ongoing. As gauged by the huge number of scientific papers on this topic, it constitutes a substantial shift in our knowledge of reproductive natural history.

REFERENCES AND FURTHER READING

Gowaty PA, Karlin AA. 1984. Multiple maternity and paternity in single broods of apparently monogamous eastern bluebirds. *Sialia sialis*. *Behav. Ecol. Sociobiol.* 15:91–95.

Jeffreys AJ, Wilson V, Thein SW. 1984. Hypervariable "minisatellite" regions in human DNA. *Nature* 314:67–73.

Avise JC. 1996. Three fundamental contributions of molecular genetics to avian ecology and evolution. *Ibis* 138:16–25.

Avise JC, Jones AG, Walker D, DeWoody JA. 2002. Genetic mating systems and reproductive natural histories of fishes: lessons for ecology and evolution. *Annu. Rev. Genet.* 36:19–45.

Oliveira RF, Taborsky M, Brockmann HJ. 2008. *Alternative Reproductive Tactics: An Integrative Approach.* Cambridge University Press, Cambridge, UK.

Chapter 59

1987
Humans Out
of Africa

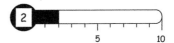

THE STANDARD PARADIGM

Anatomically modern Homo sapiens *had multiple evolutionary origins in different regions of the world.* One extreme version of this standard scenario, known as the "candelabra model", posited that different regional hominid populations had been completely isolated from one another for much more than one million years. A less extreme view held by many anthropologists was the "regional continuity" model, which supposed that modern humans arose consonantly in several Old World hominid populations that were connected by persistent gene flow across the past 1.5 million years. Much of the apparent support for these conventional ideas had come from evolutionary interpretations of (quite limited) fossil evidence.

THE CONCEPTUAL REVOLUTION

Based mostly on nucleotide sequence comparisons of maternally inherited mtDNA from modern peoples around the world, scientists proposed a recent "out-of-Africa" or "African replacement" model in which modern humans purportedly arose in Africa within the last 200,000 years and then spread throughout the Old World by replacing populations of *Homo erectus* or archaic *H. sapiens*. The mitochondrial data generally were interpreted as supporting the African replacement model because the mtDNA genealogy displayed a remarkably shallow global depth, and the inferred root of the gene tree appeared to be in Africa. In the popular press, this new phylogeographic worldview (see Chapter 51) soon became known as the "Garden of Eden" molecular hypothesis and our shared matrilineal ancestor was dubbed "mitochondrial Eve".

J.C. Avise: Conceptual Breakthroughs in Evolutionary Genetics.
DOI: http://dx.doi.org/10.1016/B978-0-12-420166-8.00059-2
125

PS-score: 2

Notwithstanding its huge impact in biological anthropology, this conceptual revolution gets only a low *PS-score* for several reasons. First, the precise timing and the number of out-of-Africa migration(s) remain highly debatable, even in this modern era when much more molecular evidence has become available. Second, the initial notion of a mitochondrial Eve sometimes was misinterpreted to imply that only a single female was alive at the time of the coalescence in the mtDNA gene tree. But, as soon was pointed out by several authors, under some plausible population demographic conditions all mtDNA lineages in existence today likely will have coalesced within the relevant timeframe even if the human population 100 millennia ago consisted of perhaps tens of thousands of unrelated females. Thus, contrary to some earlier interpretations, the existence of a mitochondrial Eve does not necessarily imply a severe population bottleneck in human numbers. Third, the out-of-Africa scenario has been challenged by some other genetic evidence suggesting extensive genetic interchange between human populations and major population expansion events that resulted in interbreeding rather than strict replacement. Finally, unlike many other conceptual breakthroughs discussed in this book, the out-of-Africa scenario applies to only a single species.

REFERENCES AND FURTHER READING

Cann RL, Stoneking M, Wilson AC. 1987. Mitochondrial DNA and human evolution. *Nature* 325:31–36.

Stringer CB, Andrews P. 1988. Genetic and fossil evidence for the origin of modern humans. *Science* 239:1263–1268.

Wolpoff MH. 1989. Multiregional evolution: the fossil alternative to Eden. In: Mellars P, Stringer C (Eds). *The Human Revolution: Behavioural and Biological Perspectives on the Origins of Modern Humans*. Edinburgh University Press, Edinburgh, UK, pp. 62–108.

Lewin R. 1993. *Human Evolution: An Illustrated Introduction*, 3rd edition. Blackwell, Oxford, UK.

Ayala FJ, Escalante A, O'Huigin C, Klein J. 1994. Molecular genetics of speciation and human origins. *Proc. Natl Acad. Sci. USA* 91:6787–6794.

Takahata N. 1995. A genetic perspective on the origin and history of humans. *Annu. Rev. Ecol. Syst.* 26:343–372.

Templeton AR. 2002. Out of Africa again and again. *Nature* 416:45–51.

Chapter 60

1989
Fossil DNA

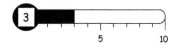

THE STANDARD PARADIGM

Substantial DNA sequences can be extracted and analyzed only from extant organisms. Except for a precious few instances in which proteins were successfully recovered from well-preserved fossil materials (such as the 40,000-year-old frozen remains of an Arctic mammoth), molecular phylogeneticists generally assumed that evolutionary trees could only be reconstructed based on macromolecular sequences obtained from living species.

THE CONCEPTUAL REVOLUTION

Beginning in the early 1980s, reports began to emerge of the successful retrieval of phylogenetically informative DNA sequences from museum-preserved or mummified material from deceased specimens of 100-year-old extinct species. These early reports involved the artificial cloning of DNA pieces in biological vectors. Then PCR (polymerase chain reaction) was invented, and this technological breakthrough appeared to change everything. In the late 1980s and early 1990s, several researchers soon claimed to have PCR-recovered long sequences of ancient DNA from the remains of organisms that had perished as long as many tens of millions of years ago. The initial flush in popularity for "fossil DNA" reached its apogee in about 1989, when an influential review of this topic was published.

PS-score: 3

Alas, many of the earliest reports now seem highly implausible. More likely, at least some of the researchers had mistakenly PCR-amplified modern DNA sequences that had contaminated their fossil material. Lengthy DNA sequences are fragile (especially under less-than-frigid environmental conditions), and today it is appreciated that a few hundred millennia is the

J.C. Avise: Conceptual Breakthroughs in Evolutionary Genetics.
DOI: http://dx.doi.org/10.1016/B978-0-12-420166-8.00060-9

maximum evolutionary age at which even the best-preserved fossil material might be expected to yield DNA sequences suitable for estimating phylogeny and molecular evolutionary rates. Thus, although fossil DNA continues to find occasional phylogenetic applications (see Chapter 66), the grandiose hopes raised by the earliest reports have not materialized to the extent originally envisioned. The history of fossil-DNA research provides a sobering reminder about the dangers of premature enthusiasm for a purported scientific breakthrough.

REFERENCES AND FURTHER READING

Lowenstein JM, Sarich VM, Richardson BJ. 1981. Albumin systematics of the extinct mammoth and Tasmanian wolf. *Nature* 291:409–411.

Higuchi R, Bowman B, Freiberger M, Ryder OA, Wilson AC. 1984. DNA sequence from the quagga, an extinct member of the horse family. *Nature* 312:282–284.

Pääbo S. 1985. Molecular cloning of ancient mummy DNA. *Nature* 314:644–645.

Pääbo S, Higuchi RG, Wilson AC. 1989. Ancient DNA and the polymerase chain reaction. *J. Biol. Chem.* 264:9709–9712.

Golenberg EM, Giannasi DE, Clegg MT, Smiley CJ, Durbin M, Henderson D, Zurawsky G. 1990. Chloroplast DNA sequence from a Miocene *Magnolia* species. *Nature* 344:656–658.

Cano RJ, Poinar HN, Pieniazek NJ, Acra A, Poinar Jr, GO. 1993. Amplification and sequencing of DNA from a 120–135-million-year-old weevil. *Nature* 363:536–538.

Lindahl T. 1993. Instability and decay of the primary structure of DNA. *Nature* 362:709–715.

Cooper A, Poinar C. 2000. Ancient DNA: Do it right or not at all. *Science* 289:1139.

Hofreiter M, Serre D, Poinar HN, Kuch M, Pääbo S. 2001. Ancient DNA. *Nature Rev. Genet.* 2:353–359.

Orlando L, Ginolhac A, Zhang G, Froese D, Albrechtsen A, Stiller M, et al., 2013. Recalibrating *Equus* evolution using the genome sequence of an early Middle Pleistocene horse. *Nature* 499:74–78.

Chapter 61

1991 Phylogenetic Character Mapping

THE STANDARD PARADIGM

Phylogenies are primarily of interest for their own sake. Despite biologists' longstanding interest in generating phylogenies from molecular and other data, additional uses for phylogenetic trees (beyond their mere reconstruction) had not yet been explored extensively.

THE CONCEPTUAL REVOLUTION

By the late 20th century, the field of molecular systematics (see Chapter 31) had grown and matured to the point where phylogenetic trees were routinely being generated for a wide diversity of taxonomic groups. Still, many phylogeneticists viewed their efforts merely as part of a larger effort to catalogue and systematize life's extraordinary diversity. The stage was set for a new worldview that would integrate molecular phylogenetic findings with the traditional kinds of phenotypic data that had always been the main source of information about organismal relationships. More specifically, molecular phylogenies could be seen to provide the genealogical framework for interpreting the precise evolutionary histories of organismal phenotypic traits. In 1991, a seminal book by Paul Harvey and Mark Pagel crystallized this sentiment into what the authors referred to as "the comparative method". In that same year, Daniel Brooks and Deborah McLennan expanded the comparative method to the field of ethology (behavior). In 2006, I christened this general method "phylogenetic character mapping". PCM involves plotting the distributions of alternative phenotypes on molecular phylogenetic trees and thereby deducing ancestral character states at internal nodes and establishing their historical patterns of evolutionary interconversion.

J.C. Avise: Conceptual Breakthroughs in Evolutionary Genetics.
DOI: http://dx.doi.org/10.1016/B978-0-12-420166-8.00061-0

PS-score: 6

The comparative method has become the *de rigueur* approach in molecular phylogenetics. Indeed, few researchers would now dispute that merely reconstructing a molecular phylogenetic tree is not the ultimate goal. If the tree is to be of any broader service, surely it must be in terms of understanding exactly how evolutionary conversions took place among particular morphological, physiological, or behavioral character states along nature's myriad evolutionary pathways.

REFERENCES AND FURTHER READING

Brooks DR, McLennan DA. 1991. *Phylogeny, Ecology, and Behavior: A Research Program in Comparative Biology*. University of Chicago Press, Chicago, IL.

Harvey PH, Pagel MD. 1991. *The Comparative Method in Evolutionary Biology*. Oxford University Press, Oxford, UK.

Harvey PH, Leigh Brown AJ, Maynard Smith J, Nee S (Eds). 1996. *New Uses for New Phylogenies*. Oxford University Press, Oxford, UK.

Avise JC. 2006. *Evolutionary Pathways in Nature*. Cambridge University Press, Cambridge, UK.

Chapter 62

1992 Evolutionary Psychology

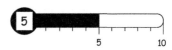

THE STANDARD PARADIGM

The human mind is a blank slate nearly free of content until written over by experience. This was a sentiment of many philosophers and scientists both before and after Darwin. In *The Descent of Man and Selection in Relation to Sex*, Darwin emphasized that the human nervous system, like any other organ system, must have evolved under the influence of natural selective forces. Nevertheless, Darwin wrote little about the human brain, preferring to leave that important task to others. More than a century would pass before the field of "evolutionary psychology" would arise to meet the challenge of trying to understand the evolved architecture of the human mind in strictly Darwinian terms.

THE CONCEPTUAL REVOLUTION

Many 20th-century authors argued that human nature and the human brain display general dispositions (such as the drive for food, shelter, and sex) consistent with selective forces operating throughout our species' long evolutionary past. Thus, the brain is not strictly a blank slate but rather it predisposes us to think and act in particular ways generally consistent with survival and reproduction.

The effective birth of evolutionary psychology in 1992 — in a landmark book by Jerome Barkow, Leda Cosmides, and John Tooby — gave added impetus to Darwinian scenarios by positing that the human brain operates much like a computer pre-loaded with content-rich neurological programs that had evolved for adaptive problem-solving in various environmental and social contexts. Under this worldview, the human mind includes a collection of modules selectively designed to solve some of the most important

J.C. Avise: Conceptual Breakthroughs in Evolutionary Genetics.
DOI: http://dx.doi.org/10.1016/B978-0-12-420166-8.00062-2

adaptive challenges encountered during humankind's evolutionary history. These behavioral modules are called upon when we find ourselves in particular predicaments. The programming might include brain circuitry adapted for particular tasks such as visual attention (what to look at in a complex scene), specific social behaviors such as various interactions with kin, or spatial navigation. For example, an evolutionary psychologist might posit that any difference in navigation skills between the sexes evidences different selective pressures on mostly male hunters versus mostly female gatherers in pre-agrarian human societies.

PS-score: 5

Evolutionary psychology has helped to transform the broader field of psychology, which traditionally entailed almost no evolutionary thinking. On the other hand, the precise degree to which specific adaptive scenarios account for the evolved architecture of the human mental machinery remains open for debate. So too do the neuronal circuits and the exact mechanistic operations of the brain modules imagined to underlie particular adaptive behaviors. Furthermore, the danger of adaptive "storytelling" (see Chapter 52) is omnipresent in evolutionary psychology because researchers can often quite easily concoct adaptive scenarios for almost any biological outcome.

REFERENCES AND FURTHER READING

Barkow JH, Cosmides L, Tooby J (Eds). 1992. *The Adapted Mind: Evolutionary Psychology and the Generation of Culture*. Oxford University Press, New York, NY.

Cela-Conde CJ, Lombardo RG, Avise JC, Ayala FJ (Eds). 2013. *In the Light of Evolution, VII: The Human Mental Machinery*. National Academies Press, Washington, DC [in press].

Cosmides L, Tooby J. 2013. Evolutionary psychology: new perspectives on cognition and motivation. *Annu. Rev. Psychol.* 64:201−229.

Striedter GF, Avise JC, Ayala FJ (Eds). 2013. *In the Light of Evolution, VI: Brain and Behavior*. National Academies Press, Washington, DC.

Chapter 63

1993
Regulatory RNAs

THE STANDARD PARADIGM

RNA molecules are involved primarily in protein translation. Standard wisdom of the era was that RNAs generally are long molecules that come in three major types: messenger (m) RNAs that are transcribed from housekeeping genes and encode polypeptides; and transfer (t) RNA molecules and ribosomal (r) RNAs that also are involved in the protein translation process on ribosomes (but see also Chapter 55).

THE CONCEPTUAL REVOLUTION

In 1993, Rosalind Lee and her colleagues discovered a class of short (each about 20 nucleotides in length) non-coding RNA molecules that proved to be regulators of eukaryotic gene expression at the transcriptional or post-transcriptional level. Known as micro (mi) RNAs, these little molecules function by base-pairing with complementary mRNA sequences from structural genes and thereby usually silencing the latter's functional expression. They have proved to be both abundant and widespread in many eukaryotic species, with the human genome (for example) housing more than 1000 miRNAs probably targeted at huge numbers of structural genes. Many miRNA genes lie in the introns of functional genes (see Chapter 49), but others reside either in exons or in intergenic regions. In effect, miRNAs probably exert their regulatory influence by turning an imprecise number of mRNA transcripts into a more precise number of protein molecules. Recently, another category of abundant non-coding RNAs − known as long non-coding RNAs or lncRNAs − has also been implicated in regulating cellular processes, perhaps by binding to the 3D structures of chromosomes. Evidence suggests that more than 10,000 lncRNAs, with roles remaining unknown, may reside in the human genome.

J.C. Avise: Conceptual Breakthroughs in Evolutionary Genetics.
DOI: http://dx.doi.org/10.1016/B978-0-12-420166-8.00063-4

PS-score: 6

To the extent that gene regulation is important in evolution and ontogeny (which now seems insuperable), the discovery of regulatory miRNAs must rank rather high on the list of conceptual breakthroughs in molecular evo-devo during the past two decades. The mode of operation of these molecules is broadly reminiscent of some earlier scenarios for how structural genes might be coordinately modulated by batteries of regulatory loci (see Chapter 41). Furthermore, some authors (e.g., Peterson et al., 2009) speculate that miRNAs have played key roles in shaping metazoan evolution.

REFERENCES AND FURTHER READING

Lee RC, Feinbaum RL, Ambros V. 1993. The *C. elegans* heterochronic gene lin-4 encodes small RNAs with antisense complementarity to lin-14. *Cell* 75:843−854.

Lau NC, Lim LP, Weinstein EG, Bartel DP. 2001. An abundant class of tiny RNAs with probable regulatory roles in *Caenorhabditis elegans*. *Science* 294:858−862.

Lee CT, Risom T, Strauss WM. 2007. Evolutionary conservation of microRNA gene complexity and conserved microRNA-target interactions through metazoan phylogeny. *DNA Cell Biol.* 26:209−218.

Rana TM. 2007. Illuminating the silence: understanding the structure and function of small RNAs. *Nat. Rev. Mol. Cell Biol.* 8:23−26.

Heimberg AM, Sempere LF, Moy VN, Donoghue PC, Peterson KJ. 2008. MicroRNAs and the advent of vertebrate morphological complexity. *Proc. Natl Acad. Sci. USA* 105:2946−2950.

Peterson KJ, Dietrich MR, McPeek MA. 2009. MicroRNAs and metazoan macroevolution: insights into canalization, complexity, and the Cambrian explosion. *BioEssays* 31:736−747.

Pennisi E. 2013. Long noncoding RNAs may alter chromosome's 3D structure. *Science* 340:910.

Chapter 64

1994
Darwinian
Medicine

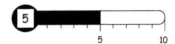

THE STANDARD PARADIGM

Darwinian thought has relatively little to offer the field of medicine. Except for some specialized genetic topics such as inborn errors of human metabolism (see Chapter 8), mother—fetal conflicts during pregnancy (see Chapter 46), and the phenomenon of senescence (see Chapter 26), the conceptual tenets of evolutionary biology seemed largely irrelevant to the actual practices of human medicine and epidemiology. Thus, evolutionary biology and Darwinian reasoning were (and are) seldom included in the advanced curricula of medical schools.

THE CONCEPTUAL REVOLUTION

In 1994, a Darwinian physician Randolph Neese teamed with the evolutionary biologist George Williams to publish a landmark book (*Why We Get Sick*) that brought widespread attention to a radically new worldview in which evolutionary perspectives would be incorporated more fully into our understanding of human illnesses and disease. "Darwinian medicine" has four general themes (Ellison et al., 2009): (1) like other organisms, humans are biological contraptions shaped by natural selection to maximize reproduction rather than health *per se*; (2) many diseases arise from a mismatch of our bodies to modern environments (because cultural change proceeds much faster than biological evolution); (3) infection is largely unavoidable (because pathogens evolve much faster than humans); and (4) many genetic variants interact during ontogeny with environmental factors and with other genes to influence human disease phenotypes. From these four foundational themes flow a wide range of conceptual and practical corollaries at the now-busy intersection of evolution, medicine, pharmacology, and epidemiology.

J.C. Avise: Conceptual Breakthroughs in Evolutionary Genetics.
DOI: http://dx.doi.org/10.1016/B978-0-12-420166-8.00064-6

PS-score: 5

Darwinian (see Chapter 1) and Hamiltonian (see Chapter 32) perspectives have the potential to transform the conceptual underpinnings (and in many cases the actual practices) of medicine and the other health sciences. However, the extent to which this potential ultimately will be realized largely remains to be seen.

REFERENCES AND FURTHER READING

Neese RM, Williams GC. 1994. *Why We Get Sick: The New Science of Darwinian Medicine.* Random House, New York, NY.

Stearns SC (Ed.). 1998. *Evolution in Health and Disease.* Oxford University Press, Oxford, UK.

Trevathan WR, McKenna JJ, Smith EO (Eds). 1999. *Evolutionary Medicine.* Oxford University Press, New York, NY.

Foster KR. 2005. Hamiltonian medicine: why the social lives of pathogens matter. *Science* 308:1269–1270.

Ellison PT, Govindaraju DR, Neese RM, Sterns SC (Eds). 2009. *Evolution in Health and Medicine. Proc. Natl Acad. Sci. USA* 107 (suppl. 1): 1691–1807.

Gluckman P, Beedle A, Hanson M. 2009. *Principles of Evolutionary Medicine.* Oxford University Press, Oxford, UK.

Chapter 65

1999
Lateral Transfer in
the Web of Life

THE STANDARD PARADIGM

The "Tree of Life" provides a proper metaphor for life's phylogenetic history. The only figure in Darwin's *On the Origin of Species* depicted a phylogenetic tree (albeit an unattractive rendition). However, the German philosopher and evolutionary biologist Ernst Haeckel did far more to make an iconography of the tree metaphor by gracing his 1866 book — *Generelle Morphologie der Organismen* — with many lovely and fully life-like arbor diagrams. Even since then, evolutionary biologists and systematists have been striving to describe the precise phylogenetic structure of the major trunks as well as multitudinous branches and twigs in the metaphorical tree of life (see Chapter 31).

THE CONCEPTUAL REVOLUTION

In the latter part of the 20th century, a consensus gradually emerged that instances of horizontal (or lateral) genetic transfer between lineages might have been so important (especially in the early history of life) as to necessitate a shift to a "web of life" scenario in which lineage anastomosis (i.e., reticulate evolution) is a major phylogenetic feature. This new paradigm was famously encapsulated in a 1999 article by Ford Doolittle when he wrote (p. 2124):

If "chimerism" or "lateral gene transfer" cannot be dismissed as trivial in extent or limited to special categories of genes, then molecular phylogeneticists will have failed to find the "true tree" not because their methods are inadequate or because they have chosen the wrong genes, but because the history of life cannot properly be represented as a tree.

J.C. Avise: Conceptual Breakthroughs in Evolutionary Genetics.
DOI: http://dx.doi.org/10.1016/B978-0-12-420166-8.00065-8

PS-score: 3

Today, there is little dispute that lateral genetic exchanges have occurred during the history of life by several mechanisms of horizontal gene transfer (HGT) such as transformation (the uptake of foreign DNA by a cell), transduction (viral mediated movement of DNA between cells), and endosymbiosis (see Chapter 38). Current scientific interest centers on the frequencies, mechanisms, and evolutionary consequences of HGT, both early in the history of life and perhaps continuing today. Gene movement between lineages also occurs quite routinely by introgressive hybridization (see Chapter 28), but this actually can be deemed a special case of vertical heredity because the DNA then passes strictly from parents to their progeny. It is also true that many scientists continue to estimate phylogenies for various taxa under the traditional paradigm that they are reconstructing branches in a mostly non-anastomatic tree. Thus, it remains to be seen whether the tree-of-life or the web-of-life metaphor ultimately prevails, both in the microbial and in the multicellular realms.

REFERENCES AND FURTHER READING

Haeckel E. 1866. *Generelle Morphologie der Organismen*. Georg Reimer, Berlin, Germany.

Doolittle WF. 1999. Phylogenetic classification and the universal tree. *Science* 284:2124–2128.

Ochman H, Lerat E, Daubin V. 2005. Examining bacterial species under the specter of gene transfer and exchange. *Proc. Natl Acad. Sci. USA* 102:6595–6599.

Arnold ML. 2006. *Evolution Through Genetic Exchange*. Oxford University Press, Oxford, UK.

Rocha EPC. 2013. With a little help from prokaryotes. *Science* 339:1154–1155.

Schönknecht G, Chen W-H, Ternes CM, Barbier GG, Shrestha RP, Stanke M., et al., 2013. Gene transfer from bacteria and archaea facilitated evolution of an extremophilic eukaryote. *Science* 339:1207–1210.

Chapter 66

2001 Genomic Sequencing

THE STANDARD PARADIGM

Complete sequencing is infeasible for the nuclear genomes of eukaryotic taxa. Ever since the molecular revolution began in earnest during the 1960s (see Chapter 37), the acquisition of extensive DNA sequences had been the Holy Grail of evolutionary genetics. In the mid-1970s, the first rapid methods for nucleotide sequencing were introduced by two research teams led by Frederick Sanger and A.R. Coulson, and Allan Maxam and Walter Gilbert. Yet even by the early 1990s it seemed almost inconceivable that complete nucleotide sequences could ever be obtained from huge eukaryotic genomes typically composed of billions of base pairs (bp).

THE CONCEPTUAL REVOLUTION

In 2001, simultaneous publications from two separate research groups stunned the scientific world by reporting complete genomic sequences of the three-billion bp human genome. This landmark achievement resulted from dramatic improvements in laboratory methods for DNA sequencing coupled with faster computer-based approaches for assembling and interpreting the extensive nucleotide sequence data. These dual technologies have been further refined in each successive iteration of "next-generation" sequencing, to the point where today it is now practicable to retrieve entire genomic sequences from almost any eukaryotic species in real time at a reasonable cost. Thus, we recently have entered the age when personalized medicine and individualized diagnostics based on complete genomic sequences are rapidly becoming scientific realities. Molecular phylogenetics and systematics (see Chapter 31), as well as the forensic (see Chapter 67) and health

J.C. Avise: Conceptual Breakthroughs in Evolutionary Genetics.
DOI: http://dx.doi.org/10.1016/B978-0-12-420166-8.00066-X

sciences (see Chapter 64), are among the many biological fields that will be greatly impacted by such ready access to voluminous genomic information. Whole-genome sequences have even been retrieved from the fossilized bones (see Chapter 60) of Neandertals and other archaic peoples (including the Denisovans) with which our ancestors apparently interbred. For example, current estimates based on whole-genome sequences suggest that about 2% of the DNA in living non-African peoples originated from interbreeding with Neandertals (close human relatives who populated Europe and western Asia from about 300,000 to 30,000 years ago).

On a related research front, laboratory methods have recently been developed to sequence whole genomes isolated from single cells (e.g., of a bacterium). This newfound capability will open many new doors not only for examining extant multicellular life but also for refined phylogenetic investigations of microbial taxa (see Chapter 50) including the many species that cannot be cultured in the laboratory.

PS-score: 7

This paradigm shift receives high marks for its scientific implications more so than for its impact to date. Indeed, as a result of successive breakthroughs in genomic sequencing, scientists now better appreciate that whole-genome sequencing is merely a starting point for more thorough scientific investigations, with the greater challenge being to annotate genomic sequences by unraveling the functional roles as well as phylogenetic origins of their myriad genetic loci.

REFERENCES AND FURTHER READING

Lander ES, International Human Genome Sequencing Consortium, et al., 2001. Initial sequencing and analysis of the human genome. *Nature* 409:860–921.

Venter JC, Adams MD, Myers EW, Li PW, Mural RJ, Sutton GG, et al., 2001. The sequence of the human genome. *Science* 291:1304–1351.

Mikkelsen TS, Hillier LW, Eichler EE, Zody MC, Jaffe DB, Yang S, et al., 2005. Initial sequence of the chimpanzee genome and comparison with the human genome. *Nature* 437:69–87.

Green RE, Krause J, Briggs AW, Maricic T, Stenzel U, Kircher M, et al., 2010. A draft sequence of the Neandertal genome. *Science* 328:710–722.

Woyke T, Tighe D, Mavromatis K, Clum A, Copeland A, Schackwitz W, et al., 2010. One bacterial cell, one complete genome. *PLoS ONE* 5:1–9.

Church G, Dudley J, Weinstock G. 2013. What's next in next-generation sequencing? *The Scientist* 27:60–61.

Pennisi E. 2013. More genomes from Denisova cave show mixing of early human groups. *Science* 340:799.

Chapter 67

2003
Barcoding Life

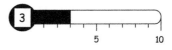

THE STANDARD PARADIGM

The main goal of DNA sequencing in an evolutionary context is to recon-
struct phylogeny. With wholesale improvements in DNA sequencing and
annotation (see Chapter 66), many researchers contemplated the prospect of
reconstructing a full phylogenetic Tree of Life for all forms of life on Earth
(see Chapter 31).

THE CONCEPTUAL REVOLUTION

In 2003, the geneticist Paul Hebert and his colleagues introduced another
perspective that focused instead on the use of DNA sequences to identify
and catalogue biological specimens. They called the approach "DNA barcoding", in obvious reference to how shopping items are barcoded in stores.
The molecule of choice for DNA barcoding became a cytochrome oxidase
gene (COI) in mitochondrial DNA. Researchers began barcoding a wide
variety of creatures, such that by early 2013 the Barcode of Life database
(http://boldsystems.org) contained nearly two million barcode sequences
from more than 160,000 species of animals, fungi, and plants.

PS-score: 3

The popularity of DNA barcoding testifies to the success of its underlying
philosophy, which is to provide a standardized approach to cataloguing life's
diversity. DNA barcoding has found applications in a wide variety of
research arenas, such as identifying: cryptic species; body parts such as avian
feathers or fish scales; leaves from unknown plants; an animal's diet based
on its stomach contents or feces; insect larvae; commercial products made
from biological materials; ancient lifeforms from fossilized material (see
Chapter 60); microbial taxa; and a host of other forensic functions. There is

J.C. Avise: Conceptual Breakthroughs in Evolutionary Genetics.
DOI: http://dx.doi.org/10.1016/B978-0-12-420166-8.00067-1

even the tantalizing prospect of developing portable hand-held devices that could quickly be used to identify samples in nature based on their DNA barcodes.

REFERENCES AND FURTHER READING

Hebert PDN, Cywinska A, Ball SL, deWaard JR. 2003. Biological identifications through DNA barcodes. *Proc. R. Soc. Lond. B* 270:313–321.

Lambert DM, Baker A, Huynen L, Haddrath O, Hebert PD, Millar CD. 2005. Is a large-scale DNA-based inventory of ancient life possible? *J. Hered.* 96:279–284.

Will KW, Mishler BD, Wheeler QD. 2005. The perils of DNA barcoding and the need for an integrative taxonomy. *Syst. Biol.* 54:844–851.

Rubinoff D. 2006. DNA barcoding evolves into the familiar. *Conserv. Biol.* 20:1548–1549.

Kress WJ, Erickson DL. 2008. DNA barcodes: genes, genomics, and bioinformatics. *Proc. Natl Acad. Sci. USA* 105:2761–2762.

Ward RD, Hanner R, Hebert PDN. 2009. The campaign to DNA barcode all fishes. *J. Fish Biol.* 74:329–356.

Chapter 68

2003
Functional
Pseudogenes

THE STANDARD PARADIGM

Pseudogenes are junk DNA. Pseudogenes traditionally were defined as genetic loci bearing close structural resemblance to known functional genes but that themselves are non-functional due to genetic alterations such as additions, deletions, or nucleotide substitutions. They represent the descendents of functional genes from which they originated via various gene duplication processes (see Chapter 39). Pseudogenes come in two basic types according to their precise mode of origin: processed pseudogenes that lack introns (see Chapter 49) because they arise when functional messenger RNA is retrotranspositionally inserted into the genome (see Chapter 42); and non-processed pseudogenes that are the evolutionary remnants of tandemly duplicated genes that became disabled by post-formational mutations. Ever since their discovery in the 1970s, pseudogenes were thought to provide quintessential examples of useless genomic junk. This intracellular "trash" also appeared to be pervasive. In the human genome, for example, about 12,000 DNA sequences were interpreted to show evidence of being pseudogenes (compared to about 25,000 functional protein-coding loci).

THE CONCEPTUAL REVOLUTION

Beginning in the 1980s, examples gradually came to light in which some "pseudogenes" played active cellular roles, such as in regulating gene expression (see Chapter 41) or generating useful genetic diversity. This new worldview – that many pseudogenes have utilitarian genomic roles – reached a crescendo in 2003 with the publication of an influential review touting the concept that pseudogenes should be considered "potogenes": DNA sequences with the evolutionary *pot*ential for becoming new functional loci.

J.C. Avise: Conceptual Breakthroughs in Evolutionary Genetics.
DOI: http://dx.doi.org/10.1016/B978-0-12-420166-8.00068-3
143

PS-score: 5

This radical new paradigm is broadly important because it provides a fine example of the concept of evolutionary preadaptation or exaptation (see Chapter 52) at the molecular level. Evolutionary geneticists now appreciate that many otherwise puzzling features of the eukaryotic genome, such as mobile elements (see Chapter 24) remain present in genomes in part because they have been coopted to serve modern adaptive cellular roles that may be entirely different from their original evolutionary *raison d'être*.

REFERENCES AND FURTHER READING

Proudfoot N. 1980. Pseudogenes. *Nature* 286:840−841.

Li W-H, Gojobori T, Nei M. 1981. Pseudogenes as a paradigm of neutral evolution. *Nature* 292:237−239.

Vanin E. 1985. Processed pseudogenes: characteristics and evolution. *Annu. Rev. Genet.* 19:253−272.

Brosius J, Gould SJ. 1992. On "genomenclature": a comprehensive (and respectful) taxonomy for pseudogenes and other "junk DNA". *Proc. Natl Acad. Sci. USA* 89:10706−10710.

Balakirev ES, Ayala FJ. 2003. Pseudogenes: are they "junk" or functional DNA? *Annu. Rev. Genet.* 37:123−151.

Torrents D, Suyama M, Zdobnov E, Bork PA. 2003. A genome-wide survey of human pseudogenes. *Genome Res.* 13:2559−2567.

Wen Y-Z, Zheng L-L, Qu L-H, Ayala FJ, Lun Z-R. 2012. Pseudogenes are not pseudo any more. *RNA Biology* 9:1−6.

Chapter 69

2004
Adaptive
Speciation

THE STANDARD PARADIGM

Allopatric speciation is the norm in sexually reproducing taxa. Ever since the mid-1900s (see Chapter 20), the prevailing worldview was that species normally arise during or immediately following geographic separation. Populations either undergo full speciation in allopatry, or they quickly complete the process by reinforcement selection for prezygotic RIBs (reproductive isolating barriers) after regaining secondary geographic contact. Strictly sympatric speciation, by contrast, was deemed to be rare or atypical at best.

THE CONCEPTUAL REVOLUTION

Beginning mostly in the 1980s, a shifting attitude emerged in which ecological and behavioral selective factors are deemed to be prime drivers of biological speciation, sometimes even in sympatry. Under this view, diversifying selection on resource utilization or mate choice frequently eventuates in reproductively isolated populations worthy of full species recognition. Some famous putative examples involve insects that rapidly shifted hosts or switched habitats, and fish species flocks that apparently arose quickly within particular bodies of water. In 2004, Ulf Dieckmann and his colleagues compiled many such examples of what they termed "adaptive speciation".

PS-score: 2

We are currently in the midst of this ongoing scientific revolution, so it is too early to know how it will unfold. In any event, it is appropriate to conclude this book by using this as the penultimate example of paradigm shifts

J.C. Avise: Conceptual Breakthroughs in Evolutionary Genetics.
DOI: http://dx.doi.org/10.1016/B978-0-12-420166-8.00069-5
145

in evolutionary genetics, because it brings us back to the Darwinian revolution (see Chapter 1) that opened this book. Darwin too envisioned that species arise by means of natural selection operating during ecological plays on the evolutionary stage. Thus, in a sense we have come conceptually full circle!

REFERENCES AND FURTHER READING

Berlocher SH, Bush GL. 1982. An electrophoretic analysis of *Rhagoletis* (Diptera: Tephritidae) phylogeny. *Syst. Zool.* 31:136–155.

West-Eberhard M. 1983. Sexual selection, social competition, and speciation. *Q. Rev. Biol.* 58:155–183.

Echelle AA, Kornfield I (Eds). 1984. *Evolution of Fish Species Flocks.* University of Maine Press, Orono, ME.

Kondrashov AS, Mina MV. 1986. Sympatric speciation: When is it possible? *Biol. J. Linn. Soc.* 27:201–223.

Schluter D. 2000. *The Ecology of Adaptive Radiation.* Oxford University Press, Oxford, UK.

Schluter D. 2001. Ecology and the origin of species. *Trends Ecol. Evol.* 16:372–380.

Dieckmann U, Doebeli M, Metz JAJ, Tautz D. (Eds). 2004. *Adaptive Speciation.* Cambridge University Press, Cambridge, UK.

Chapter 70

2010
Comparative
Genomics

THE STANDARD PARADIGM

Genomic-level data are difficult to obtain but then should be rather straight-forward to interpret. Prior to the turn of the 21st century it seemed unthinkable that whole-genome molecular data would become commonplace, so most research effort involved developing laboratory procedures for gathering DNA sequences from focal taxa such as humans or fruit flies (see Chapter 66). Thus, the field of evolutionary genomics focused on a few special-interest organisms toward which extraordinary sequencing effort was expended.

THE CONCEPTUAL REVOLUTION

Genome-level data became relatively easy to obtain but remain surprisingly difficult to interpret. After the past decade, following several successive technical advances in "next-generation" sequencing, we now find ourselves — for the first time — in an era when genome-wide data can be obtained from multiple individuals of any species (model or otherwise). In other words, several iterations of next-generation sequencing have allowed genomes to be surveyed in multiple individuals or populations rather than merely in a few focal specimens. Thus we now find ourselves at the early dawn of a modern age in which we are limited not by available genomic data but rather by our capacity to analyze and annotate the vast new stores of genomic information.

PS-score: 8

It is too soon to guess how this ongoing revolution in population genomics will play out, but it seems fair to surmise that nearly every topic addressed

J.C. Avise: Conceptual Breakthroughs in Evolutionary Genetics.
DOI: http://dx.doi.org/10.1016/B978-0-12-420166-8.00070-1
147

in this book eventually will merit re-examination from a population geno-mics perspective. We have already learned, for example, that reproductive barriers separating species can be semipermeable to genetic exchange, with some components of the genome flowing across species' barriers far more freely than others; that particular genetic adaptations sometimes arise conver-gently during the evolutionary process; that natural selection and genetic drift differentially impact different segments of the genome; that evolution-ary clocks tick at very different paces in different genomic regions; and that the genealogical histories of unlinked genomic segments vary considerably from locus to locus in sexual species. In addition to providing genomic sequences *per se*, the genomic revolution is also supporting the rise of related fields such as transcriptomics (the study of expressed genes and their transcribed RNAs), proteomics (the large-scale study of proteins), metabolo-mics (the large-scale study of metabolic networks), canceromics (the study of cancers), and other "omics" disciplines relevant to developmental biology, genetics, and evolutionary biology.

This book has offered an aerial survey of conceptual breakthroughs in evolutionary genetics during the 150 years since Darwin. It is interesting to contemplate what the next century (or even the next few decades) will bring. Will the pace of key discoveries — perhaps stimulated by the ongoing geno-mics revolution — continue to accelerate, or is it possible that the era of fun-damental paradigm shifts in evolutionary genetics has already passed its peak, such that much of what remains will entail relatively minor refine-ments or adjustments of current thought? If history is a guide, the fields of evolutionary biology and genetics will someday look vastly different from now, but that will be a subject for future historians of science to address.

REFERENCES AND FURTHER READING

Wu CI. 2001. The genic view of the process of speciation. *J. Evol. Biol.* 14:851−865.

Pollard KS, Salama SR, Lambert N, Lambot MA, Coppens S, Pedersen JS., et al., 2006. An RNA gene expressed during cortical development evolved rapidly in humans. *Nature* 443:167−172.

Avise JC. 2010. Conservation genetics enters the genomics era. *Conserv. Genet.* 11:14456−14459.

Burke MK, Dunham JP, Shahrestani P, Thornton KR, Rose MR, Long AD. 2010. Genome-wide analysis of a long-term evolution experiment with Drosophila. *Nature* 467:587−590.

Hohenhole PA, Bassham S, Etter PD, Stiffler N, Johnson WA, Cresko WA. 2010. Population genomics of parallel adaptation in threespine stickleback using sequenced RAD tags. *PloS Genetics* 6:e1000862.

Narum SR, Buerkle A, Davey JW, Miller MR, Hohenlohe PA. 2013. Genotyping-by-sequencing in ecological and conservation genomics. *Molec. Ecol.* 22:2841−2847.

Epilogue

The scientific revolutions described in this book represent only the tip of a vast iceberg of challenges to conventional wisdom in evolutionary genetics. Nearly every published book or journal article in the field tries to make the case that its author has uncovered some new conceptual truth that flies in the face of traditional thought. Thus, every evolutionary geneticist is a visionary at some level, for, if not, he or she would not possess the pioneering spirit that ultimately drives all scientific research. Although only a small fraction of research discoveries achieves the status of major paradigm busters, this fact should not dissuade those of us who work daily in the scientific trenches, because most scientific revolutions were quite unanticipated by their architects. If this book has accomplished nothing else, I hope that it has achieved two overarching objectives: (1) to remind students of the relatively recent and sometimes tenuous origins of evolutionary genetic concepts that were visionary for their era but now might seem so obvious as to be nearly self-evident truths; and (2) to open readers' eyes to the endless supply of scientific paradigms that continually warrant critical re-examination. That is, after all, how science progresses.

Adaptation Any feature (e.g., morphological, physiological, behavioral) that makes an organism suited to survive and reproduce in a particular environment.

Aging (See senescence.)

Allele Any of the possible alternative forms of a gene. A diploid individual carries two alleles at each autosomal gene, and these can either be identical in state (in which case the individual is homozygous) or different in state (heterozygous). At each autosomal gene, a population of N diploid individuals harbors $2N$ alleles, many of which may differ somewhat in nucleotide sequence.

Allopatric Occurring in separate geographic areas.

Alternative splicing The ligation of exons of a particular gene to form a functional RNA that differs in information content from the normal messenger RNA.

Altruism Self-harmful behavior performed for the benefit of others.

Amino acid One of the molecular subunits polymerized to form polypeptides.

Antagonistic pleiotropy A form of pleiotropy in which a gene's influence on one phenotypic trait is beneficial to the organism but its influence on another trait is detrimental.

Artificial selection Selection that operates in plant or animal populations through the human choice of particular genetically-based traits.

Autosome A chromosome in the nucleus other than a sex chromosome; in diploid organisms, autosomes are present in homologous pairs. (See also sex chromosome.)

Bacterium A unicellular microorganism without a true cellular nucleus.

Biochemistry The chemistry of life.

Biogeography The study of the geographical distributions of organisms.

Cancer A disease characterized by uncontrolled cellular proliferation.

Cell A small, membrane-bound unit of life capable of self-reproduction.

Central dogma The outdated notion that genetic information invariably flows from DNA to RNA to proteins.

Chemostat An apparatus that allows the continuous cultivation of bacterial populations in a closed and controlled environment.

Chimera An individual composed of a mixture of genetically different cells.

Chloroplast The chlorophyll-containing, photosynthesizing organelle of plants.

Chromatin The complex of DNA and proteins of which eukaryotic chromosomes are composed.

Chromosome A threadlike structure within a cell that carries genes.

Clade A monophyletic assemblage; a group of species (or, sometimes, conspecific individuals) that share a closer common ancestry with one another than with any other such group.

Cladistics A method of classification that seeks to reconstruct phylogenies in terms of branching patterns of ancestral-descendent lineages.

Clone A group of genetically identical cells or organisms, all descended from a single ancestral cell or organism; or, the process of creating such genetically identical cells or organisms.

Codominance A genetic situation in which both alleles in a heterozygous diploid individual are expressed simultaneously in the phenotype.

Coevolution The joint evolution of two or more ecologically interacting species.

Complementary DNA (cDNA) DNA produced from an RNA template by a reversal of transcription.

Conspecific Belonging to the same species.

Cryptic female choice A phenomenon by which the female reproductive tract influences which sperm will fertilize her eggs.

Cytoplasm The portion of a eukaryotic cell outside of the nucleus.

Deleterious In reference to genetics, any genetic condition that harms an organism's health.

Demography The study of the dynamics of populations.

Deoxyribonucleic acid (DNA) A double-stranded molecule each of whose nucleotide subunits is composed of a deoxyribose sugar, a phosphate group, and one of the nitrogenous bases adenine, guanine, cytosine, or thymine.

Diploid A usual condition of a somatic cell in which two copies of each chromosome are present. (See also haploid.)

Dominant allele An allele that, in a heterozygous diploid individual, is expressed fully in the phenotype. (See also recessive allele.)

Drift (See genetic drift.)

Duplication A genetic event usually stemming from an abnormal meiosis in which a gene or portion of a chromosome gives rise to a second copy.

Ecology The study of the relationships between organisms and their biotic and physical environments.

Ecosystem A community of organisms and the physical environment with which it interacts.

Egg dumping A phenomenon in which a female lays one or more eggs in the nest of another female.

Endosymbiosis A form of symbiosis in which one organism lives within the body of another.

Enzyme A protein that catalyzes a specific chemical reaction.

Epidemiology The study of disease epidemics, with an emphasis on tracing the causes.

Epigamic selection A type of sexual selection that operates via female choice of mates.

Epigenetics Changes in gene expression that are inherited but not caused by changes in DNA sequence *per se*. In a broader sense, epigenetics can also refer to the entire suite of mechanisms, developmental pathways, and social and other environmental influences by which genomes give rise to organismal-level features.

Epistasis A genetic situation in which two or more genes influence one another's expression in the production of a phenotype.

Eugenics The ideology or practice of attempting to improve *Homo sapiens* by altering its genetic composition.

Eukaryote Any organism in which chromosomes are housed in a membrane-bound nucleus.

Eusocial A highly social system (such as of many bees, wasps, and ants) characterized by cooperative care of offspring and reproductive divisions of labor.

Evolution Change through time in the genetic composition of a population.

Exaptation A biological adaptation in which the current biological function of a trait in question originally evolved for reasons unrelated to that trait's current role.

Exon A gene segment that codes for a polypeptide. (See also intron.)

Fitness (genetic) The contribution of a genotype to the next generation relative to the contributions of other genotypes in the population.

Fossil Any remain or trace of life no longer alive.

Fraternal twins Siblings derived from separate zygotes within a pregnancy.

Game theory The study of strategic decision-making in behavioral interactions.

Gamete A mature reproductive sex cell (egg or sperm).

Gametogenesis The specialized series of cellular divisions that leads to the production of gametes. (See also meiosis, oogenesis, spermatogenesis.)

Gene The basic unit of heredity; usually taken to imply a sequence of nucleotides specifying production of a polypeptide or other functional product such as ribosomal RNA, but also can be applied to stretches of DNA with unknown or unspecified function.

Gene flow The movement of genes between populations.

Gene pool The sum total of all hereditary material in a population or species.

Genealogy A record of descent from ancestors through a pedigree.

Genetic code The consecutive nucleotide triplets of DNA and RNA that specify particular amino acids for protein synthesis.

Genetic drift Change in allele frequency in a finite population by chance sampling of gametes from generation to generation.

Genetic engineering Any experimental or industrial method employed to alter the genomes of living cells.

Genetic load The collective burden of genetic defects in a population.

Genetics The study of heredity and hereditary molecules.

Genome The complete genetic constitution of an organism; can also refer to a particular composite piece of DNA, such as the mitochondrial genome.

Genotype The genetic constitution of an individual organism with reference to a single gene or set of genes. (See also phenotype.)

Germ cell A sex cell or gamete. (See also somatic cell.)

Germline Pertaining to the cellular lineage from which germ cells are derived.

Gerontology The scientific study of the processs of growing old.

Group selection Selection that operates on multiple members of a hereditary lineage as a unit.

Haploid A usual condition of a gametic cell in which only one copy of each chromosome is present. (See also diploid.)

Heredity Genetic inheritance; the phenomenon of familial transmission of genetic material from generation to generation.

Heritability The fraction of variation of a trait within a population due to heredity as opposed to environmental influences.

Heterosis The condition in which heterozygotes have higher genetic fitness than homozygotes.

Heterozygote A diploid organism possessing two different alleles at a specified gene. (See also homozygote.)

Homeotic gene A gene that controls the overall body plan of an organism by influencing the developmental fate of groups of cells.

Homology Similarity of structure due to inheritance from a shared ancestor; can refer to any structural features ranging from DNA sequences to morphological traits.

Homoplasy Parallel or convergent evolution; structural similarity in organisms not due to inheritance from a shared ancestor.

Homozygote A diploid organism possessing two identical alleles at a specified gene. (See also heterozygote.)

Hormone An organic compound produced in one region of an organism and transported to target cells in other parts of the body where its effects on phenotype are exerted.

Hybridization The mating of individuals belonging to genetically disparate populations or species.

Inbreeding The mating of kin.

Inbreeding depression Lowered genetic fitness due to the mating of kin.

Inclusive fitness The sum of an individual's personal genetic fitness plus that individual's influences on genetic fitness of relatives other than direct descendents.

Independent assortment (Mendel's Law of) The random distribution to gametes of alleles from genes on different chromosomes, or from genes far enough apart on a given chromosome.

Intelligent Design (ID) A recent incarnation of the Creation Science movement in which a supreme intelligence is invoked to account for biological complexity and perfection.

Introgression Hybridization-mediated movement of genes between species.

Intron A non-coding portion of a gene. Most genes in eukaryotes consist of alternating intron and exon DNA sequences.

Jumping gene (See transposable element.)

Junk DNA DNA that contributes nothing beneficial to the organism in which it is housed.

Karyotype The somatic chromosomal complement of an individual or species, often shown in a diagrammatic format.

Kin selection A form of natural selection due to individuals favoring or disfavoring the survival and reproduction of genetic relatives other than offspring.

Lamarckian inheritance A largely discredited notion of heredity involving the transmission of genetic attributes gained during the lifetime of a parent.

Linkage disequilibrium Non-random allelic associations across loci on the same chromosome.

Linkage, genetic The co-occurrence of particular loci on the same chromosome, thus often implying (if the loci are close enough) some restriction on recombination between them.

Locus (pl. loci) A gene, or a specified stretch of DNA.

Macroevolution Long-term evolution above the taxonomic level of species.

Macromolecule A large biological molecule such as a nucleic acid or protein.

Maternal inheritance Heredity via the female parent only.

Meiosis The cellular process whereby a diploid cell divides to form haploid gametes.

Meiotic drive Any mechanism in meiosis that results in the unequal recovery of the two types of gametes produced by a heterozygote.

Messenger RNA A form of ribonucleic acid transcribed from structural genes, the exon-derived portions of which subsequently will be translated into a polypeptide.

Metabolic pathway A series of stepwise biochemical changes in the conversion of a precursor substance to an end product, each step typically catalyzed by a specific enzyme.

Metabolism The sum of all physical and chemical processes by which living matter is produced and maintained, and by which cellular energy is made available.

Metabolomics The large-scale study of metabolism.

Microbe An organism too small to be seen with the unaided eye.

Microevolution Short-term evolution within a species.

Microorganism (See microbe.)

Micro-RNA (miRNA) A short stretch of RNA that can bind to complementary sequences in a messenger RNA molecule and thereby inhibit the translation or induce the degradation of a specific genetic message.

Microsatellite locus A stretch of DNA containing short repeated sequences each typically about two to six base-pairs in length.

Minisatellite locus A stretch of DNA containing medium-length repeated sequences each typically about 10 to 70 base pairs in length.

Mitochondrial Of or pertaining to the mitochondrion.

Mitochondrion An organelle in the cell cytoplasm that contains its own DNA (mtDNA) and that is the site of some of the metabolic pathways involved in cellular energy production.

Mitosis A process of cell division that produces daughter cells with the same chromosomal constitution as the parent cell. (See also meiosis.)

Mobile element (See transposable element.)

Molecular clock The notion that genetic loci tend to evolve at a stochastically constant pace.

Monogamy A mating system in which each individual has only one mate.

Monozygotic twins Genetically identical siblings that trace back to a single fertilized egg within a pregnancy.

Morphogenesis The emergence of an organism's morphological appearance during ontogeny.

Mutagenic Mutation-inducing.

Mutation A spontaneous change in the genetic constitution of an organism.

Natural selection The differential survival and reproduction of individuals with different genotypes.

Natural Theology The idea that the beauties and workings of nature provide final and definitive proof of God's majesty.

Neutral mutation A mutation that neither enhances nor diminishes genetic fitness.

Nucleic acid (See deoxyribonucleic acid or ribonucleic acid.)

Nucleotide A unit of DNA or RNA consisting of a nitrogenous base, a pentose sugar, and a phosphate group.

Nucleus The portion of a eukaryotic cell bounded by a nuclear membrane and containing chromosomes.

Ontogeny The development of an individual from fertilized egg to maturity.

Oocyte An egg cell.

Oogenesis The production of oocytes (i.e. unfertilized eggs).

Organ A recognizable body feature (such as the heart or kidney) composed of several different tissues grouped together into a functional and structural unit.

Organelle A complex cytoplasmic structure of characteristic morphology and function, such as a mitochondrion or plastid.

Paleontology The study of fossils and the evolutionary history of now-extinct life.

Paradigm A conceptual framework within which scientific theories are constructed.

Paralogy Structural similarity of replicate genes due to their common origin via a gene duplication event.

Parasite An organism that lives on or in an organism of a different species, and derives nutrients from its host.

Parsimony Economy or sparingness of scientific explanation for a phenomenon.

Phenotype The observable properties of an organism at any level, ranging from molecular and physiological to gross morphological.

Phylogeny Evolutionary relationships (historical descent) of a group of organisms or species.

Phylogeography The study of the spatial distributions of genealogical lineages within and among related species.

Phytophagous Plant eating.

Plastid A self-replicating cytoplasmic organelle of plant cells. (See also chloroplast.)

Pleiotropy A genetic phenomenon wherein a single gene or genetic alteration influences multiple phenotypic features.

Pluripotency Pertaining to embryonic or other cells that retain the capacity to generate multiple types of tissues or organs.

Polyandry A mating system in which females have multiple mates.

Polygamy A mating system in which an individual has multiple mates.

Polygenic trait A phenotypic trait affected by multiple genes.

Polygynandry A mating system in which both males and females have multiple mates.

Polygyny A mating system in which males have multiple mates.

Polymer A macromolecule composed of a bonded collection of repeating subunits or monomers.

Polymerase An enzyme that catalyzes the formation of nucleic acid molecules.

Polymorphism With respect to genetics, the presence of two or more genotypes in a population.

Polypeptide A polymer composed of amino acids chemically linked together.

Population bottleneck A severe but often temporary reduction in the size of a population.

Population genetics The study of evolutionary forces that can change the genetic composition of populations.

Preadaptation (See exaptation.)

Prokaryote Any microorganism that lacks a chromosome-containing, membrane-bound nucleus.

Protein A biological macromolecule composed of one or more polypeptide chains.

Protein electrophoresis A laboratory method by which proteins are separated in a matrix based on their electrical charges.

Proteomics The large-scale study of proteins.

Protozoan A unicellular animal.

Pseudogene A gene bearing close structural resemblance to a known functional gene at another chromosomal site, but that itself is non-functional due to genetic alterations such as additions, deletions, or nucleotide substitutions.

Quantitative trait A phenotype that is quantitative in nature and typically shows a continuous distribution in a population.

Recessive allele An allele that, in a heterozygous diploid individual, is masked in phenotypic expression by a dominant allele at the same locus. (See also dominant allele.)

Recombinant DNA A new DNA molecule that has arisen from genetic recombination (often mediated by humans).

Recombination (genetic) The formation of new combinations of genes through such natural processes as meiosis and fertilization, or in the laboratory through recombinant DNA technologies.

Regulatory gene A gene that exerts operational control over the expression of other genes.

Repetitive DNA A DNA sequence that is present in multiple copies within a genome.

Retrotransposable element A mobile element that moves about the genome via an intermediate RNA molecule which then is reverse-transcribed into DNA.

Retrovirus An RNA virus that utilizes reverse transcription during its life cycle to integrate into the DNA of a host cell.

Reverse transcriptase An enzyme that catalyzes the conversion of an RNA sequence into a corresponding DNA sequence.

Ribonucleic acid (RNA) A single-stranded polynucleotide molecule each of whose nucleotide subunits is composed of a ribose sugar, a phosphate group, and one of the nitrogenous bases adenine, guanine, cytosine, or uracil.

Ribosomal RNA A form of ribonucleic acid that together with ribosomal proteins composes a ribosome.

Ribosome An organelle in the cell cytoplasm composed of RNA and protein and that is the site of protein translation.

Ribozyme An RNA sequence with enzymatic capability.

Science Objective knowledge, or the means by which such knowledge is attained, through careful observation, experiment, and logical inference.

Segregation (Mendel's Law of) The distribution to gametes of the two alleles in a diploid individual; each gamete receives, at random, one or the other of the two alleles at each gene.

Selfish DNA DNA that displays self-perpetuating modes of behavior without apparent benefit to the organism.

Senescence A persistent decline in the age-specific survival probability or reproductive output of an individual due to internal physiological deterioration.

Sex chromosome A chromosome in the cell nucleus involved in distinguishing the two genders. In humans, the "X" and "Y" are sex chromosomes. (See also autosome.)

Sex ratio The relative number of males and females in a population.

Sexual reproduction Reproduction involving the production and subsequent fusion of haploid gametes.

Sexual selection The differential ability of individuals of the two genders to acquire mates. The topic can be subdivided into two components: intrasexual selection, which refers to competition among members of the same sex over access to mates; and intersexual or epigamic selection, which refers to mating choices made between males and females.

Sociobiology The study of the biology of social interactions among individuals.

Soma (See somatic cell.)

Somatic cell Any cell in a eukaryotic organism other than those destined to become germ cells. (See also germ cell.)

Speciation The process by which new species arise.

Species A group of actually or potentially interbreeding individuals that is reproductively isolated from other such groups.

Sperm competition A phenomenon whereby sperm cells compete against one another, often within a female's reproductive tract, for fertilization of an egg.

Spermatogenesis The production of sperm.

Spliceosome A large ribonucleoprotein that biochemically removes all intron-derived segments from each pre-mRNA and then splices each gene's exons end-to-end to generate a mature mRNA.

Spontaneous generation A sudden emergence of life from non-living material.

Stochastic Chancy; governed largely or exclusively by random events.

Symbiosis A close interaction or association (usually implied to be mutually beneficial) between organisms belonging to different species.

Sympatric Occurring in the same geographic area.

Symplesiomorphy An ancestral trait that is shared by two or more descendent taxa.

Synapomorphy A derived trait that is shared by two or more taxa by virtue of direct inheritance from a shared ancestor.

Syngamy The union of two gametes to produce a zygote; fertilization.

Systematics The study of evolution and classification of organisms into a hierarchical series of groups.

Taxonomy The theory and practice of naming and classifying organisms.

Theodicy Vindication of the justice of God in establishing a world in which evil exists.

Theology The study of God, religion, and revelation.

Tissue A biological feature composed of similar cells organized into a functional and structural unit.

Transcription The cellular process by which an RNA molecule is formed from a DNA template.

Transcriptomics The large-scale study of expressed genes and their transcribed RNAs.

Transduction Viral-mediated movement of DNA between cells.

Transfer RNA A form of ribonucleic acid that picks up amino acids from the cell cytoplasm and moves them into position for the translation process.

Transformation The uptake of foreign DNA by a cell.

Transgenic Carrying genetic material from another organism or species.

Translation The process by which the genetic information in messenger RNA is employed by a cell to direct the construction of polypeptides.

Transposable element Any of a class of DNA sequences that can move from one chromosomal site to another, often replicatively.

Transposition The process by which a replica of a transposable element is inserted into another chromosomal site.

Transposon A kind of transposable element that is immediately flanked by inverted and direct repeats.

Virus A tiny, obligate intracellular parasite incapable of autonomous replication but which instead utilizes the host cell's replicative machinery.

X-chromosome The sex chromosome normally present as two copies in female mammals and insects (the homogametic sex), but as only one copy in males (the heterogametic sex).

Y-chromosome In mammals and many other species, the sex chromosome normally present in males only.

Zygote The diploid cell arising from the union of male and female haploid gametes.

Index

Printed and bound by CPI Group (UK) Ltd, Croydon, CR0 4YY

03/10/2024

01040421-0015